HISTORY OF ORIENTAL ASTRONOMY

HISTORY OF ORIENTAL ASTRONOMY

Proceedings of an International Astronomical Union
Colloquium No. 91
New Delhi, India
13–16 November 1985

Edited by

G. SWARUP

A. K. BAG

K. S. SHUKLA

CAMBRIDGE UNIVERSITY PRESS
Cambridge
New York New Rochelle Melbourne Sydney

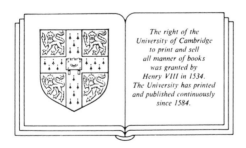

Published by the Press Syndicate of the University of Cambridge
The Pitt Building, Trumpington Street, Cambridge CB2 1RP
32 East 57th Street, New York, NY 10022, USA
10 Stamford Road, Oakleigh, Melbourne 3166, Australia

First published 1987

Printed in Great Britain at the University Press, Cambridge

British Library cataloguing in publication data

The History of oriental astronomy:
proceedings of the International Astronomical Union,
colloquium no. 91, New Delhi, India,
13 November to 16 November 1985.
1. Astronomy – Asia – History
I. Swarup, G. II. Bag, A. K. III. Shukla, K. S.
520′.95 QB33.A78

Library of Congress cataloguing in publication data

History of oriental astronomy.
Includes indexes.
1. Astronomy, Hindu - History - Congresses.
2. Astronomy, Chinese - History - Congresses.
3. Astronomy, Arabic - History - Congresses.
I. Swarup, G. (Govind), 1929- . II. Bag, A. K.
III. Shukla, Kripa Shankar. IV. International Astronomical Union.
V. Title: Oriental astronomy.
QB18.H58 1987 520′.9 87-18376

ISBN 0 521 34659 2

Contents

Preface

IAU Colloquium 91 on History of Oriental Astronomy was held at the Indian National Science Academy, New Delhi, from November 13 to November 16, 1985. Eighty four participants from nineteen countries including forty eight from India attended the symposium. The Colloquium was sponsored by the International Astronomical Union (IAU); International Union for the History and Philosophy of Science; Council of Scientific & Industrial Research,New Delhi; Department of Science & Technology through the National Organizing Committee for the IAU General Assembly,New Delhi; Indian Institute of Astrophysics,Bangalore; Indian National Science Academy (INSA),New Delhi; Tata Institute of Fundamental Research (TIFR),Bombay and University Grants Commission, New Delhi.

Forty six papers were presented at the Colloquium of which ten were invited papers, covering mainly the characteristics and achievements of oriental astronomy;inter-regional development and mutual influences; ancient data relating to eclipses, supernovae, comets; mediaeval astronomical developments; instruments; observatories and related topics with emphasis on how theoretical studies were gradually nourished by the observational astronomy. All the contributed papers presented in the Colloquium were critically examined by the editors and as such, some of the papers have not been included in these Proceedings. We are grateful to Professor E.S.Kennedy and Dr.Michel Teboul for editing some of the manuscripts. The papers in the Proceedings have been arranged broadly in chronological order of the era they refer to.

Several social and cultural events, including an evening of Indian classical dances, were organised during the Colloquium to give the participants a touch of India. Visits were arranged to the well known Jai Singh Observatory in New Delhi built of stone, the famous Qutub Minar, the ancient Iron Pillar and also to Son-et Luminiere at the Red Fort which depicted the history of India over the last 500 years, particularly of the Mughal period.

An exhibition of charts giving astronomical contributions of the oriental cultures, photographs of scientific instruments from earlier stages to the seventeenth century and display of important Indian manuscripts and source books were organized which attracted a great deal of interest among the participants.

A book on **History of Astronomy in India** edited by S.N.Sen and K.S. Shukla, published by INSA, New Delhi, was released to commemorate the occasion.

We are grateful to many colleagues in INSA and TIFR for their valuable contributions to the successful organisation of the Colloquium. We are specially thankful to Mrs. Sabnam Shukla and Mr.B.N.Chakrabarty of INSA for their valuable help. Only eight out of the thirty-eight manuscripts included in these Proceedings were provided by the authors on camera-ready sheets. We are thankful to Mrs.Mary Helen Kennedy for final typing of five papers and to Mr.M.Mathews, Mr.C.Ramu and Ms. N.N. Shanthakumary for their meticulous typing of all the other manuscripts in these Proceedings. Although we have carefully compared these with the original manuscripts, we regret for any errors that might have crept in. We thank Mr.C.Gangeyamoorthy for tracing the figures and to Mr.C.Ramachandra Rao for the photographic work.

SCIENTIFIC ORGANISING COMMITTEE

Ansari, S.M.R. (India)
Kennedy, E.S. (FRG) (Chairman)
King, D. (USA)
Mercier, R. (UK)
Pedersen, O. (Denmark)
Pingree, D. (USA)
Saliba, G. (Lebanon)
Xi Zezong (China)
Yabuuti, K. (Japan)

LOCAL ORGANISING COMMITTEE

Ansari, S.M.R.
Bag, A.K. (Secretary)
Jain, Ashok
Kothari, D.S.
Mathur, B.S.
Rahman, A.
Shukla, K.S.
Singh, Col. J.E.S.
Subbarayappa, B.V.
Swarup, G. (Chairman)

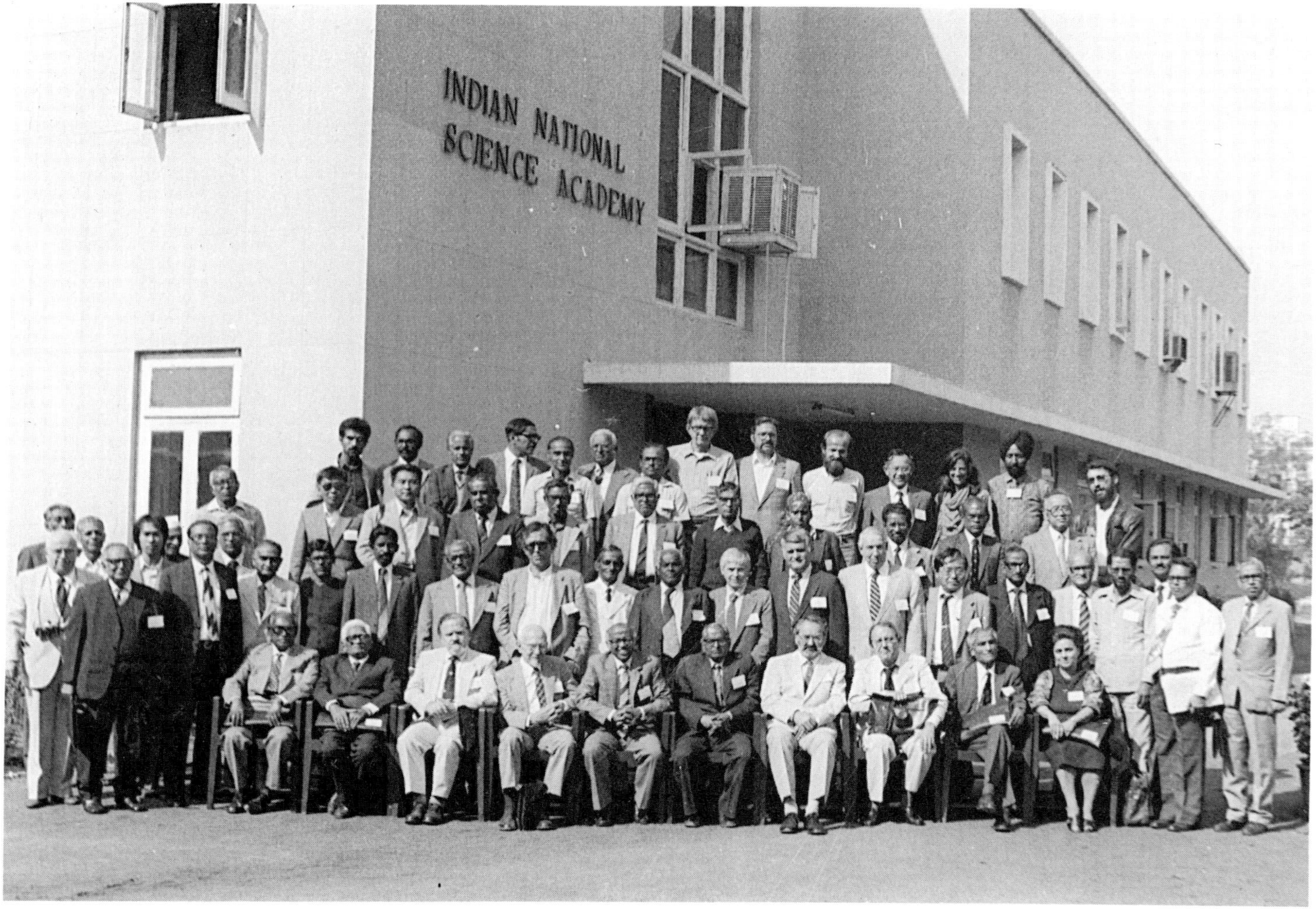
INDIAN NATIONAL
SCIENCE ACADEMY

1. S.N.Sen
2. F.C.Auluck
3. O.Pedersen
4. E.S.Kennedy
5. G.Swarup
6. K.S.Shukla
7. R.Mercier
8. Paul Kunitzsch
9. S.K.Chatterjee
10. Maria Firneis
11. A.F.Karl Fischer
12. George Abraham
13. L.C.Jain
14. S.B.Roy
15. Y.Ohashi
16. M.B.Pant
17. A.K.Bag
18. A.K.Chakravarty
19. A.N.Mohammed Fadhl
20. Parameswar Jha
21. Sahban Ahmad
22. S.K.Mukherjee
23. J.Samso
24. M.C.Chaki
25. K.V.Sarma
26. O.Gingerich
27. J.A.Eddy
28. J.E.Kennedy
29. Michio Yano
30. R.C.Gupta
31. J.E.S. Singh
32. Mohammad Ilyas
33. A.K.Bhatnagar
34. Ramaratnam
35. L.S.Kothari
36. R.R. Kalvala
37. Quan Hejun
38. Xu Zhen-tao
39. T.R.Chandrasekhar
40. Omkar Nath
41. J.N.Nanda
42. S.Parthasarathy
43. S.D.Sharma
44. S.R.Sarma
45. A.K.Das
46. S.Nakayama
47. S.M.Hadi Hadavi
48. T.Velusamy
49. V.R.Venugopal
50. S.M.R.Ansari
51. A.Bandyopadhyay
52. W.H.Abdi
53. Ramatosh Sarkar
54. W.C. Livingston
55. J. Hamadani-Zadeh
56. Gene Ammarell
57. Bambang Hidayat
58. J.Saad-Cook
59. S.S.Lishk
60. Y.T. Langerman

LIST OF PARTICIPANTS

ABRAHAM,G., 1520, 12th Main Road, Annanagar, Madras 600 040, India.

AMMARELL,G., Education Director, University of Colorado, Fiske, Planetarium and Science Centre, Campus Box 408, Boulder, Colorado 80309, U.S.A.

ANNA MANI, Indian Institute of Tropical Meteorology, Field Research Unit, Wind Energy Survey Project, 239 A Rajmahal Vilas Extension, Bangalore 560 080, India.

ANSARI,S.M.R., Department of History of Medicine & Science, Institute of History of Medicine & Medical Research, P.O. Hamdard Nagar, New Delhi 110 062, India.

AULUCK,F.C., 5 Vaishali, Pitampura, New Delhi 110 034, India.

BAG,A.K., Head, History of Science Division, Indian National Science Academy, Bahadur Shah Zafar Marg, New Delhi 110 002, India.

BANDYOPADHYAYA,A., Director, Positional Astronomy Centre, P-546, Block-N, New Alipore, Calcutta 700 053, India.

BHATNAGAR,A., Positional Astronomy Centre, P-546, Block-N, New Alipore, Calcutta 700 053, India.

BHATTACHARYYA,J.C., Director, Indian Institute of Astrophysics, Bangalore 560 034, India.

BOGGINO,A.E.T., Institute de Ciencias, Basicas, Universidad Nacional De Asuncion, Casillas de Carreo 1039, Asuncion, Paraguay.

CHAKI,M.C., 27, Sashi Bhusan De Street, Calcutta 700 012, India.

CHAKRAVARTY,A.K., Department of Mathematics, M.R. College, Mahishadal 721 628, Midnapore, West Bengal, India.

CHANDRASEKHAR,T.R., Member, Administrative Staff College of India, V-13, Green Park Extension, New Delhi 110 016, India.

CHANDRATREYA,G.L., 'Seetaram', 12 33-B-Apte Road, Pune 411 004, India.

CHATTERJEE,S.K., A-9/1 Vasant Vihar, New Delhi 110 057, India.

CHEN MEIDONG, 43 Instructor, Inst. for the History of Natural Science, Academia Sinica, Beijing, China.

DESAI,P.M., Vedhashala Astronomical Observatories, Navrangpura Railway Crossing, Ahmedabad 380 013, India.

EDDY,J.A., NCAR, Box 3000 Boulder, Colarado 80307, U.S.A.

FADHL,A.N.M., Director,National Observatory Project, P.O. Box 6086, Riyadh 1144, Saudi Arabia.

FIRNEIS,M., Astronomical Institute, University of Vienna, Turkenschanzstr. 17, A-1180 Vienna, Austria.

FISCHER,A.F.K., No.84 ROTT, F-67, 160 Wissembourge, France.

GINGERICH,O., Harvard-Smithsonian Center for Astrophysics, Churchill College, Cambridge, CB3-ODS,U.K.

GOVIND,V., Raja Ram Mohan Roy National Education Research Centre, IW 3-Curzen Road Barracks, Kasturba Gandhi Marg, New Delhi 110 001, India.

GUPTA,R.C., Professor of Mathematics, Birla Institute of Technology, P.O. Mesra, Ranchi, India.

HABIB,I., National Institute of Science, Technology and Development Studies, Hillside Road, New Delhi 110 012, India.

List of participants

HADAAVI,S.M.H., Physics Department, College of Science, Shiraz University, Shiraz-71454, Iran.

HAMADANI-ZADEH,J., Department of Mathematics and Computer Science, Sharif University of Technology, P.O.Box 11365-8939, Tehran, Iran.

HIDAYAT,B., Director, Boscha Observatory, Lembeng, Java, Indonesia .

HUGONNARD-ROCHE,H., 9 Rue Spontini, 75116 Paris, France.

ILYAS,M., School of Physics, University of Science of Malayasia, Penang, Malayasia.

IWANISZEWSKA,V.,vice-President, IAU Commission-46, Institute of Astronomy, N.Copernicus University, Torun, Poland.

JAIN, ANUPAM, Assistant Professor of Mathematics, Govt.Degree College, Biaora, Rajgarh (M.P.) 465 674, India.

JAIN, ASHOK., Director, National Institute of Science, Technology and Development Studies, Hillside Road, New Delhi 110 012, India.

JAIN,L.C., 'Surya Emporium', 677 Sarafa, Dist.Jabalpur, Madhya Pradesh, India.

JANI,B.H., Vedhshala Astronomical Observatories, Navrangpura Railway Crossing, Ahmedabad 380 013, India.

JANI,H.M., Vedhshala Astronomical Observatories, Navrangpura Railway Crossing, Ahmedabad 380 013, India.

KALVALA,R.R., CSIR Pool Officer, Centre for Advanced Studies in Astronomy, Osmania University, Hyderabad 500 007, India.

KENNEDY,E.S., Institut fur Geschichte, Der Arabisch-Islamischen Wissenschaften, Beethovenstr, 32, D-6000, Frankfurt 1, Germany.

KENNEDY,J.E., 323 Lake Cres, Saskatoon, Sask, D, S7II 3AI, Canada.

KHAN,M.S., Park Street,Post Box No.9072, Calcutta-700 016, India.

KHANNA,A., National Institute of Science, Technology and Development Studies, Hillside Road, New Delhi 110 012, India.

KOCHHAR,R.K., Indian Institute of Astrophysics, Bangalore 560 034, India.

KOKOTT,W., Postfach-255, 8012 Ottobrun, Germany.

KUNITZSCH,P., Davidstrasse 17, D-8000 Munchen 81, Germany.

LANGERMAN,Y.T., Faculty of Humanities, Institute of Asian and African Studies,The Hebrew University of Jerusalem, Jerusalem, Israel.

LEROG,E., Doggett, U.S.A.

LISHK,S.S., Lecturer (Mathematics), Government in-Service Training Centre, Patiala 147 001, India.

LIVINGSTON,W.C., U.S.A.

MATHUR,B.S., National Physical Laboratory, New Delhi 110 012, India.

MERCIER,R., Faculty of Mathematical Studies, University of Southampton, S 09 5 NH, Great Britain, U.K.

MITRA,A.P., Director General, Council of Scientific and Industrial Research, Rafi Marg, New Delhi 110 001, India.

MITRA,S., Indian National Science Academy, 1, Park Street, Calcutta 700 016, India.

List of participants

MUKHERJEE,S.K., Vice-Chairman, National Commission for the Compilation of History of Sciences in India, 332, Jodhpur Park, Calcutta 700 068, India.

NAKAYAMA,S., College of General Education, University of Tokyo, Komba, Meguru, Tokyo, Japan.

NANDA,J.N., 181, Indira Nagar Colony, Dehradun-248 009, India.

OHASHI, Y., 3-5-26 Hiroo, Shibuya-Ku, Tokyo, Japan.

OTAROD,S., Physics Department, Razi University, Bakhtaran, Iran.

PANT,M.B., 58, Narayan Peth, Pune 411 030, India.

PARTHASARATHI,S., National Physical Laboratory, Hillside Road, New Delhi 110 012, India.

PEDERSEN,O., History of Science Department, Aarhus University, 8000 Aarhus C, Denmark.

QUAN HEJUN, 51 Instructor, Shanghai Observatory, Academia Sinica, Shanghai, China.

RAINA,D., National Institute of Science Technology and Development Studies, Hillside Road, New Delhi 110 012, India.

ROY,S.B., Director, Institute of Chronology, B-20 Sujan Singh Park, New Delhi 110 003, India.

SAAD-COOK,J., 1204 South Washington Street, Apartment 817 W, Alexandria, Virginia 22314, USA.

SAMSO,J., Calle San Pedro, 1-3-5-7 Gasa 11, San Cugat de Valles, Barcelona, Spain.

SARKAR,R., Birla Planetarium, 96, Chowringhee Road, Calcutta 700 071, India.

SARMA,K.V., 55, Govinda Vilas, Theosophical Society, Adyar, Madras, 600 020, India.

SARMA,S.R., Department of Sanskrit, Aligarh Muslim University, Aligarh, India.

SEN,S.N., Rama Krishna Mission Institute of Culture, Gol Park, Ballygunge, Calcutta 700 029, India.

SHARMA,S.D., P-16, Panjabi University Campus, Patiala, Punjab, India.

SHARMA,V.N., Department of Physics & Astronomy, University of Wisconsin, Fox Valley Centre, Menasha, W1 54952-1297, USA.

SHUKLA,K.S., 'Argara', Hussainganj Crossing, Behind Bata Shoe Co., Lucknow 226 001, India.

SIDHARTHA,B.G., Birla Planetarium, Naubat Puhad, Adarsh Nagar, Hyderabad 500 463, India.

SINGH,J.E.S., Director, Nehru Planetarium, Teen Murti Marg, New Delhi 110 011, India.

SINGH,N., Scientist, National Institute of Science, Technology & Development Studies, Hillside Road, New Delhi 110 012, India.

SWARUP,G., Senior Professor, Tata Institute of Fundamental Research, Indian Institute of Science Campus, Post Box 1234, Bangalore 560 012, India.

TEBOUL,M., C.N.R.S. & Tokyo Astronomical Observatory, The University of Tokyo, Tokyo, Japan.

TEKELI,S., Din ve Turih-Cografya Fakultesi, Turk Kulturinu Arastirmo, Fustitusu, Ankara, Turkey.

List of participants

VELUSAMY,T., Radio Astronomy Centre (TIFR), P.O.Box No.8, Udhagamandalam 643 001, India.

VENUGOPAL,V.R., Radio Astronomy Centre (TIFR), P.O.Box No.8, Udhagamandalam 643 001, India.

XI ZEZONG, Institute for the History of Natural Science, Academia Sinica, Beijing, China.

XU ZHEN-TAO, 42 Instructor, Purple Mountain Observatory, Academia Sinica, Nanjing, China.

YABUUTI,K., 20, Tanaka, Higasi-Hinakutityo, Sakyo-Ku, Kyoto, Japan.

YANO,M., International Institute for Linguistic Sciences, Kyoto Sangyo University, Kamigamo, Kita-Ku, Kyoto (603), Japan.

1
Introductory lecture

"Let me close with an oversimplified geographical recapitulation. We find our subject originating a few centuries before the Christian era in two disparate cultures, Mesopotamia and the Hellenistic world. From the Mediterranean it passed to India, there to flourish. Thence the centroid of activity moved westward, residing in the lands of Islam during medieval times, more recently in Europe. Now astronomical research is carried out throughout the entire world."

- E.S. Kennedy (p.6)

E.S.Kennedy examining books displayed at the astronomical exhibition which was organized during the Colloquium.

1 INTRODUCTORY LECTURE

E.S. Kennedy
Institut für Geschichte der Arabisch-Islamischen Wissenschaften, Beethovenstr,32 D-6000 Frankfurt 1, West Germany

When I undertook to serve as Chairman of the Programme Committee for this meeting, I was consoled with the reflection that,in my capacity as Chairman,I would be able to arrange that I should not make a speech and I took steps along these lines. But upon my arrival in Delhi, I was presented with a programme on which my name appeared as speaker. Well, so be it. I will attempt a survey of ancient and mediaeval astronomy.

In the programme, the title says Oriental astronomy. But if we delete (which we should not) the fundamental Hellenistic contribution, then we can forget the adjective. We can say that ancient and mediaeval astronomy was all oriental. Now, in essaying these remarks, I am subject to at least two contraints, the first being clearly that I may not speak except about the branches of the subject of which I have some personal knowledge. And the consequence is that I can say nothing about Chinese astronomy, which is a pity, and for which I apologise to the colleagues from China. And secondly, from birth I have been endowed with an extremely bad memory, and as my age increases, the ability to forget things quickly likewise increases. Therefore, the picture which I will try to paint perforce must be done with an extremely broad brush, and with the omission of practically all details. And so I begin.

In the Bible, it says, "In the beginning was the Word". For astronomy then I say, "In the beginning was the Number". By that I mean that a necessary condition for the existence of astronomy is that the prospective astronomer be in a position to carry out long and complicated computations. That is to say, it is essential for the existence of astronomy,that there exist a place-value number system. I mean, an arrangement whereby, given any real number, that number may be represented to any desired degree of precision by a series of integer multiples of integer powers of a fixed integer, the base. The invention of the first place-value number system has been said to be of equal importance with the invention of alphabetical writing, and I think justly so. Being in possession of an alphabet, the user may represent any word whatsoever by the use of a finite number of different symbols, just as with a place-value number system any number is representable by the use of a finite number of different symbols. I remind you that the thing of importance is neither the existence of the base, nor what particular integer happens to be the base, but rather the fact that in the array which represents the number, the contribution of any particular digit - any particular multiplier - to the value of its number is determined not only by the symbol itself, but also by its position in the array. The place-value system was invented in Mesopotamia shall we say, some 3000 years ago. The base was the number sixty, and hence this is called the sexagesimal system. Sometime after that, but, I suppose, well before the beginning of the Christian era, a second place-value number system was invented, here in India, with base ten. It has been claimed that the notion of place-value travelled from ancient Mesopotamia to India. Be that as it may, I think no one contests the fact that the decimal system came out of India. Having done so, it travelled slowly westward, penetrating the lands where the sexagesimal system was used, Western Asia, North Africa and Europe. However, at no time did the decimal system for astronomical computations replace the sexagesimals. But it should be stated that the system as used for astronomical computations was a mixed arrangement involving two bases. By and large, the numbers were small, measuring arcs on a circle, hence less than 360 degrees. So it was not inconvenient to represent the integer part of a number with a non-place-value decimal, only the fractional part being displayed as a place value sexagesimal. It should also be admitted that frequently, since computations in the decimal system are simpler than those with sexagesimals, many astronomers had the habit of converting into decimals a pair of numbers to be multiplied, performing the operation, and then turning the product back into sexagesimals.

As for astronomy proper, we can say the subject originated more or less simultaneously in two places, one of which was Mesopotamia (modern Iraq) in the fifth century B.C. The Babylonian astronomers, being long in possession of the sexagesimal system, developed extremely powerful methods for calculating the true positions of planets. Apparently behind this technique there was no geometrical model. The approach was purely numerical, and the prime tools of these astronomers were step functions and linear zig-zag functions. This extraordinary body of theory, however, soon vanished, disappearing completely until modern times, when clay tablets bearing calculations in cuneiform script were recovered with the excavation of ancient sites, and eventually deciphered.

At more or less the same time, in the eastern basin of the Mediterranean, the Greeks worked out a second variety of solutions to the same fundamental problem, the computation of planetary positions, by means of geometric models which we may put into two categories. On the one hand, the celestial body was assumed to be rotating in a circular orbit with a constant speed, but with the earth fixed not at the centre of the orbit, but displaced away from it. Hence this was called the eccentric hypothesis. The second arrangement retains the earth at the centre of the circular orbit, but the point rotating about the earth had in turn, rotating about it a satellite, as it were, the actual planet. The two rotations then proceeded simultaniously. This is known as the epicyclic hypothesis.

These notions made their way to India by means, I suppose, of maritime commerce merchants plying between the southwest coast of India and the Red Sea thence by land to the Nile delta. The Indians apparently had a low opinion of the eccentric hypothesis. Rather they siezed upon the epicyclic device. It was necessary to introduce two independent perturbations to the planets' motion, the anomalistic equation and the equation of center, both periodic but of different periods. The Indian astronomers obtained these effects by introducing two epicycles per planet, each with its proper period and radius. The mutual influences of each equation on the other were obtained by sequences of complicated but ingenious numerical operations.

Meanwhile, in the Mediterranean region, astronomical activity continued. In the second century A.D. Ptolemy, the greatest astronomer of antiquity, succeeded in combining the epicyclic and and eccentric approaches to produce a set of planetary models which were extremely successful, and which maintained themselves supreme for centuries, eventually being displaced only by the elliptical orbits of Kepler and a cosmology based on the Newtonian laws of motion.

Returning to the Middle East, we note that there converged upon the scientific milieu of the Abbasid empire twin influences, from the east from India (with certain Iranian modifications about which we know very little), and from the west the Ptolemaic. There ensued a competition between the two. The Indian system was the first in the field, but eventually the Ptolemaic triumphed over the Indian, the latter curiously enough surviving for a long time in Arab Spain.

In both systems, in order to obtain numerical results from the geometric models, there was a need for some variety of trigonometry, or the equivalent thereof. In Hellenistic times the need was met by a discipline which can hardly be called trigonometry, since the basic configuration involved was not the triangle but the complete plane or spherical quadrilateral: four arbitrary straight lines or great circles. To this was applied the Menelaos theorem, and for computation a single numerical table, the table of chords, sufficed. Subsequently, the Indians discarded the chord function in favour of the much handier sine, the fundamental periodic function of modern science. Knowledge of the sine function passed westward, together with information about the tangent function, evolved perhaps simultaneously both in India and the Middle East from sets of primitive shadow tables. Then the Muslims rapidly invented the other standard trigonometric functions, and developed trigonometry proper as an independent branch of mathematics which was applied primarily to the solution of problems in astronomy.

At the same time, there occurred in the Middle East an intense and sustained burst of activity in observational astronomy, and this period, say between the eighth and the fourteenth centuries, saw the emergence of the astronomical observatory as a scientific institution*. Moreover, this activity was accompanied by an unprecedented development of very sophisticated computational mathematics. For instance, there was extensive application of convergent iterative algorisms. The result of this was a proliferation of numerical function tables, again practically all of them being of interest to astronomers. An example is the sine table of Ulugh Beg, produced late in the period, in which sines were calculated for every minute of every degree from zero to ninety, precise to four sexagesimal places. That is to say, to a precision of one over sixty to the fourth power, equivalent to seven decimal places.

Finally, brief mention may be made of a theoretical development not dependent upon observation, but rather in deference to an old doctrine probably first enunciated by the Pythagoreans, which one may call the principle of uniform circularity. According to this notion, any acceptable celestial motion must consist of a combination of uniform circular motions. The Ptolemaic planetary models violate this principle, and this fact had bothered people over a long period of time. There was a succession of Muslim astronomers in Iran and in Syria who succeeded in working out planetary models which on the one hand conformed to the principle of uniform circularity, but on the other hand, yielded essentially the same results as the Ptolemaic models. Curiously enough, some time later many identically the same planetary models appeared again in the work of Copernicus, the sole difference being that whereas the Muslim models were strictly geostatic, in those of Copernicus, the earth was given an orbit.

Well, I think, this will have to suffice, and I add only a general observation which is implicit in practically everything I said, and which I now make explicitly. That is to note the symbiotic relationship which has flourished throughout history between the sister disciplines of astronomy and mathematics; each of the two subjects nurtured the other. Let me close with an oversimplified geographical recapitulation. We find our subject originating a few centuries before the Christian era in two disparate cultures, Mesopotamia and the Hellenistic world. From the Mediterranean it passed to India, there to flourish. Thence the centroid of activity moved westward, residing in the lands of Islam during medieval times, more recently in Europe. Now astronomical research is carried out throughout the entire world.

* The utility of the Symposium, at least to the speaker, is demonstrated by the false statement made above. Only during the proceedings did I learn that observatories in China antedated all others.

2
Ancient astronomy and its characteristics

In the post-vedic period the scope of astronomy was widened. Astronomy outgrew its original purpose of providing a calendar to serve the needs of the vedic priests and was no longer confined to the study of the Sun and Moon. The study of the five planets was also included within its scope and it began to be studied as a science for its own sake.

- K.S.Shukla (p.14)

"Mishra Yantra" in the Jantar Mantar Observatory at Delhi built by Jai Singh in 1724

2.1 MAIN CHARACTERISTICS AND ACHIEVEMENTS OF ANCIENT INDIAN ASTRONOMY IN HISTORICAL PERSPECTIVE

Kripa Shankar Shukla
Hussainganj Crossing, Lucknow-226 001, India

INTRODUCTION

Ancient Indian astronomy may be classified into two main categories: (1) the vedic astronomy and (2) the post vedic astronomy. The vedic astronomy is the astronomy of the vedic period i.e. the astronomy found in the vedic *saṃhitās* and *brāhmaṇas* and allied literature. The principal avocation of the people in the vedic times being the performance of the vedic sacrifices at the times prescribed by the *śāstras*, it was necessary to have accurate knowledge of the science of time so that the times prescribed for performing the various vedic sacrifices could be correctly predicted well in advance. Astronomy in those times, therefore, was essentially the science of time-determination. It centred round the Sun and Moon and its aim was to study the natural divisions of time caused by the motion of the Sun and Moon, such as days, months, seasons and years, special attention being paid to the study of the times of occurrence of new moons, full moons, equinoxes and solstices.

VEDIC ASTRONOMY

The *Ṛgveda* (1.52.11; 10.90.14), which is believed to be the earliest of the *Vedas*, describes the universe as infinite and made up of the Earth, the atmosphere and the sky. According to the *Taittiriya-saṃhitā (7.5.23), fire* rests in the Earth, the air in the atmosphere, the Sun in the sky and the Moon in the company of the *nakṣatras* (zodiacal stargroups). The *Ṛgveda* (1.105.10; 4.50.4; 10.123.1; also see *Śatapatha-brāhmaṇa*, 4.2.1) refers to the five planets as gods and mentions Bṛhaspati[1] (Jupiter) and Vena (Venus) by name[2]. It also mentions the thirty-four lights which, in all probability, are the Sun, the Moon, the five planets and the twenty-seven *nakṣatras (Ṛgveda*, 10.55.3).

The *Ṛgveda* (8.58.2; 1.95.3; 8.58.2; 1.164.14) describes the Sun as the sole lightgiver of the universe, the cause of the seasons, the controller and lord of the world (*Aitareya-brāhmaṇa* 2.7 describes sun as the cause of wind). The Moon is called *Sūrya-ras'mi* i.e. one which shines by sunlight (*Taittirīya-saṃhitā* 3.4.7.1). The Moon's path was divided into 27 equal parts, because the Moon took about 27 1/3 days in traversing it. These parts as well as the stars lying in their neighbourhood were called *nakṣatras* and given the names *Kṛttikā* etc. When the constellation called Abhijit (Lyra) was included in the list of *nakṣatras*, their number was stated as 28. Of these *nakṣatras*, *tiṣya* (or *Puṣya*), Aghā (or Maghā), Arjunī (or Phalgunī), Citrā and Revatī are mentioned in the *Ṛgveda* (5.54.13; 10.64.8.; 10.85.13; 4.51.2; 4.51.4). The *Taittirīya-saṃhitā* (4.4.10.1-3; see also *Atharva-saṃhitā*, 19.7.2-5; *Kāṭhaka-saṃhitā*, 39.13; *Maitrāyaṇī-saṃhitā*, 2.13.20) and the *Taittirīya-brāhmaṇa* (1.5.1; 3.1.1-2; 3.1.4-5) give the names of the 28 *nakṣatras* along with those of the deities supposed to preside over them. The *Śatapatha-brāhmaṇa* (10.5.4.5) gives the names of the 27 *nakṣatras* as well as those of the 27 *upa-nakṣatras*. The *nakṣatras* were categorized into male, female and neuter as well as into singular, dual and plural. It seems that the prominent stars of each *nakṣatra* were counted and classified in order of their brilliance.

Some constellations other than the *nakṣatras* were also known. The *Ṛgveda* (1.24.10; 10.14.11; 10.63.10) mentions the *Ṛkṣas* or Bears (the Great Bear and the Little Bear), the two divine Dogs (Canis Major and Canis Minor), and the heavenly Boat (Argo Navis). The Great Bear was also known as *Saptarṣi* (the constellation of the seven sages) and was mentioned by this name in the *Śatapatha-brāhmaṇa*[3] (2.1.2.4) and the *Tāṇḍya-brāhmaṇa* (1.5.5). The golden Boat (Argo Navis) is mentioned in the *Atharva-veda* (5.4.4; 6.95.2) also. The *Aitareya-brāhmaṇa* (13.9) mentions the constellation of *Mṛga* or Deer (Orion) and the star *Mṛgavyādha* (Sirius), and narrates an interesting story regarding them.

Besides the Sun, the Moon, and the *nakṣatras*, mention is also made of some of the other heavenly bodies and heavenly phenomena. For example, *ulkā* (meteors) and *dhūmaketu* (comets) have been mentioned in the *Atharvaveda*(19.9.8-9, 19.9.10). Eclipses have been mentioned and described as caused by Svarbhānu or Rāhu. The *Ṛgveda* (5.40.5-9) describes an eclipse of the Sun as brought about by Svarbhānu. The *Tāṇḍya-brāhmaṇa* (4.5.2; 4.6.13: 6.6.8; 14.11.14-15; 23.16.2) mentions eclipses as many as five times. Eclipses have been mentioned in the *Atharva-veda* (19.9.10), the *Gopatha-brāhmaṇa* (8.19) and the *Śatapatha-brāhmaṇa* (5.3.2.2) also.

The day, called *vāsara* or *ahan* in the vedic literature, was reckoned from sunrise to sunrise. The variability of its length was known. The *Ṛgveda* (8.48.7) invoking Somarāja says: "O Somarāja, prolong thou our lives just as the Sun increases the length of the days." Six days were taken to form a *ṣaḍaha* (six-day week); 5 *ṣaḍahas*, a month; and 12 months, a year. As to the names of the six days of a *ṣaḍaha*, there is no reference in the vedic literature. However, the six-day week was later replaced by the present seven day week (*saptāha*) which had attained popularity and was in general use at the time of composition of the *Atharva-jyautiṣa*.

The duration of daylight, reckoned from sunrise to sunset, was divided into two parts called *pūrvāhṇa* (forenoon) and *aparāhṇa* (afternoon), three parts called *pūrvāhṇa*, *madhyāhna*, and *aparāhṇa*, four parts called *pūrvāhṇa*,*madhyāhna*, *aparāhṇa* and *sāyāhna*,[4] and five parts called *prātaḥ, saṅgava, madhyāhna, aparāhṇa* and *sāyāhna* (*Śatapatha-brāhmaṇa*, 2.2.3.9). The days and nights were also divided into 15 parts each, and these parts were called *muhūrta*. The *muhūrtas* falling during the days of the light and dark fortnights as well as those falling during the nights of the light and dark fortnights were given specific names (*Taittirīya-brā hmaṇa* 3.10.1.1-3). The fifteen days and nights of the light fortnight as well as the fifteen days and nights of the dark fortnight were also assigned special names, (*Taittirīya-brāhmaṇa* 3.10.1.1-3; 3.10.10.2).

On the analogy of a civil day, a lunar day was also sometimes reckoned from one moonrise to the next and the name *tithi* was given to it (*Aitareya-brāhmaṇa*, 32.10). The use of the term *tithi* in the sense in which it is used now occurs in the *Vedāṅga-jyautiṣa*. (*Ārca-jyautiṣa*, 20,21,31; *Yājuṣa-jyautiṣa* 20-23 , 25, 26). It does not occur in the vedic *saṁhitās* and *brāhmaṇas*, but there are reasons to believe that *tithis* were used even in those times.

The year, generally called by the terms *samā, vatsara* and *hāyana* in the vedic literature, was seasonal or tropical and was measured from one winter solstice to the next, but in due course it was used in the sense of a sidereal year. In the early stages, therefore, the names of the seasons were used as synonyms of a year. The *Kauṣītaki-brāhmaṇa* (19.3) gives an interesting account of how the year-long sacrifice was commenced at one winter solstice and continued until the next winter solstice: "On the new moon of Māgha he (the Sun) rests, being about to turn northwards. They (the priests) also rest, being about to sacrifice with the introductory Atirātra. Thus, for the first time, they (the priests) obtain him (the Sun). On him they lay hold with the Caturviṃśa rite; that is why the laying hold rite has that name. He (the Sun) goes north for six months; him they (the priests) follow with six day rites in continuation. Having gone north for six months, he (the Sun) stands still, being about to turn southwards. They (the priests) also rest, being about to sacrifice with the Viṣuvanta (summer solstice) day. Thus, for the second time, they obtain him (the Sun). He (the Sun) goes south for six months; they (the priests) follow him with six day rites in reverse order. Having gone south for six months, he (the Sun) stands still, being about to turn north; and they (the priests) also rest, being about to sacrifice with the Mahāvrata day. Thus they (the priests) obtain him (the Sun) for the third time".

The *Taittirīya-brāhmaṇa* (3.9.22) calls the year "the day of the gods", the gods being supposed to reside at the north pole.

The year was supposed to consist of six seasons and each season of two (solar) months. The relation between the seasons and months was as shown in the following Table:

Vedic seasons (*Taitt.-saṃhitā*; 4.3.2;5.6.23;7.5.14) and months (*Taittirīya-saṃhitā*, 1.4.14;4.4.11)

Seasons	Months
1. Vasanta (Spring)	1. Madhu
	2. Mādhava
2. Grīṣma (Summer)	3. Śukra
	4. Śuci
3. Varṣā (Rainy)	5. Nabhas
	6. Nabhasya
4. Śarada (Autumn)	7. Iṣa
	8. Ūrja
5. Hemanta (Winter)	9. Sahas
	10. Sahasya
6. Śiśira (Chilly Winter)	11. Tapas
	12. Tapasya

Two (solar) months commencing with the winter solstice were called Śiśira; the next two months, Vasanta; and so on. Sometimes Śiśira and Hemanta were treated as one season and the number of seasons was taken as five (*Aitareya-brāhmaṇa*, 1.1; *Taittīrīya-brāhmaṇa*, 2.7.10).

The lunar or synodic month was measured from full moon to full moon or from new moon to new moon (*Taittirīya-saṃhitā*, 7.5.6.1) as is the case even now. The names Caitra etc. based on the *nakṣatras* in which the Moon becomes full do not occur in the early *saṃhitās* and *brāhmaṇas* but such terms as *phalgunī-pūrṇamāsī*, *citrā-pūrṇamāsī*, etc. are found to occur in the *Taittirīya-saṃhitā* (7.4.8). They occur in the *Śāṅkhāyana* and *Tāṇḍya brāhmaṇas*, the *Vedāṅga-jyautiṣa* and the *kalpa-sūtras*[5]. Twelve lunar months constituted a lunar year. In order to preserve correspondence between lunar and solar years, intercalary months were inserted at regular intervals. Mention of the intercalary month is made in the *Ṛgveda* (1.25.8), but how it was arrived at and where in the scheme of months it was introduced in that time is not known. The *Vedāṅga-jyautiṣa* prescribes insertion of an intercalary month after every 30 lunar months (*Yājuṣa-jyautiṣa*, 37). Thus a year sometimes contained 12 lunar months and sometimes 13 lunar months. The *Taittirīya-saṃhitā* (5.6.7) refers to 12 as well as 13 months of a year and calls the thirteenth (intercalary) month by the names *saṃsarpa* and *aṃhaspati* (1.4.14). The Vājasaneyi-saṃhitā (7.30;22.31) calls the intercalary month on one occasion by the name *aṃhasaspati* and on another by the name *malimluca* (22.30). In later works the synodic month with two *saṃkrāntis* is called *aṃhaspati*, the synodic month without any *saṃkrānti*, occurring before it, is called *saṃsarpa*, and the synodic month without any *saṃkrānti* occurring after it is called *adhimāsa* (intercalary month, *Tantrasaṃgraha* i.8)

Originally the lunar (or synodic) months Caitra etc. were named after the *nakṣatras* occupied by the Moon at the time of full moon. But in due course they were linked with the solar months. Thus the lunar month (reckoned from one new moon to the next) in which the Sun entered the sign Aries was called Caitra or Madhu; that in which the Sun entered the sign Taurus was called Vaiśākha or Mādhava; and so on. The lunar month in which the Sun did not enter a new sign was treated as an intercalary month.

Periods bigger than a year are also met with in the vedic literature. They were called *yuga*. One such yuga consisted of 5 solar years. The five constituent years of this *yuga* were called saṃvatsara, parivatsara, idāvatsara, anuvatsara and idvatsara. The *Ṛgveda* (7.103.7-8) mentions two of these, viz. *saṃvatsara and parivatsara*. The *Taittirīya-saṃhitā* (5.5.7.1-3), the *Vājasaneyi-saṃhitā* (27.45;30.16) and the *Taittirīya-brāhmaṇa* (3.4.11; 3.10.4), mention all the five names, with some alteration. The *Taittirīya-saṃhitā* calls them saṃvatsara, parivatsara, idāvatsara, iduvatsara and vatsara; the *Vājasaneyi-saṃhitā*, saṃvatsara, parivatsara, idāvatsara, idvatsara and vatsara and the *Taittirīya-brāhmaṇa*, saṃvatsara, parivatsara, idāvatsara, idvatsara and vatsara respectively. The names Kṛta, Tretā, Dvāpara and Kali which are used in later astronomy as the names of longer *yugas* are also used in the vedic literature to indicate different grades, each inferior to the preceeding. But Dvāpara, as a unit of time, is found to be used in the *Gopatha-brāhmaṇa* (1.1.28).

The earliest work which exclusively deals with vedic astronomy is the *Vedāṅga-jyautiṣa*. It is available in two recensions, *Ārca-jyautiṣa* and *Yājuṣa-jyautiṣa*. Both the recensions are essentially the same; a majority of the verses occurring in them being identical. The date of this work is controversal, but the situation of the Sun and Moon at the beginning of the *yuga* of five years mentioned in this work, according to T.S.Kuppanna Sastry, existed about B.C. 1150 or about B.C. 1370, according as the first point of *nakṣatra Śraviṣṭhā* stated there means the first point of the *nakṣatra*-segment *Śraviṣṭhā* or the *nakṣatra*-group *Śraviṣṭhā* (Sastry 1984,**3**,p.13). This work defines *jyotiṣa* (astronomy) as the science of time-determination and deals with months, years, muhūrtas, rising *nakṣatras*, new moons, full moons, days, seasons and solstices. It states rules to determine the *nakṣatra* occupied by the Sun or Moon, the time of the Sun's or Moon's entry into a *nakṣatra*, the duration of the Sun's or Moon's stay in a *nakṣatra*, the number of new moons or full moons that occurred since the beginning of the *yuga*, the position of the Sun or Moon at the end of a new moon or full moon day or *tithi*, and similar other things. It gives also the measure of the water-clock, which was used to measure time, and tells when an intercalary month was to be added or a *tithi* was to be omitted. In short, it gives all necessary information needed by the vedic priest to predict times for the vedic sacrifices and other religious observances.

The five-year *yuga* of the *Vedāṅga-jyautiṣa* contained 61 civil, 62 lunar and 67 sidereal months. The year consisted of 366 civil days which were reckoned from sunrise to sunrise. After every thirty lunar months one intercalary month was inserted to bring about concordance between solar and lunar years. Similarly to equate the number of *tithis* and civil days in the *yuga* of five solar years, the thirty full moon *tithis* which ended between sunrise and midday were omitted. There were six seasons of equal duration in every year, each new season beginning after every 61 days. Besides *tithis* and *nakṣatras*, the *yoga* called Vyatipāta was also in use.

The five-year *yuga* was taken to commence at the winter solstice occurring at the beginning of the first *tithi* of the light half of the month Māgha. Since the Sun and Moon were supposed to occupy the same position at the beginning of each subsequent *yuga* and all happenings in one *yuga* were supposed to be repeated in the subsequent *yugas* in the same way, the calendar constructed on the basis of the *Vedāṅga-jyautiṣa* was meant to serve for a long time.

The *Vedāṅga-jyautiṣa* astronomy suffered from two main defects. Since there are actually 1826.2819 days in a *yuga* of five solar (sidereal) years and not 1830 as stated in the *Vedāṅga-jyautiṣa*, therefore if one *yuga* was taken to commence at a winter solstice the next one commenced about four days later than the next winter solstice and not at the next winter solstice. Similarly, since there are actually 1830.8961 days in a period of 62 lunar months and not 1830 as stated in the *Vedāṅga-jyautiṣa*, therefore there was a deficit of about one *tithi* in the *yuga* of five solar years. These discrepancies must have been rectified but we do not know when and how this was done.

There is one more work on *jyotiṣa* belonging to the later vedic period It is known as *Atharva-jyautiṣa*. This work describes the *muhūrtas*, *tithis*, *karaṇas*, *nakṣatras* and week days and prescribes the deeds that should be performed in them. The names of the lords of the week days stated in this work viz. Āditya (Sun), Soma (Moon), Bhauma (the son of Earth), Bṛhaspati, Bhārgava (the son of Bhṛgu) and Śanaiścara (the slow-moving planet) are undoubtedly of Indian origin and must have been in use in India from very early times[6].

POST-VEDIC ASTRONOMY

In the post-vedic period the scope of astronomy was widened. Astronomy outgrew its original purpose of providing a calendar to serve the needs of the vedic priests and was no longer confined to the study of the Sun and Moon. The study of the five planets was also included within its scope and it began to be studied as a science for its own sake. While further improvement of luni-solar astronomy continued, astronomers now devoted their attention towards the study of the planets which were known in the vedic period and were now well known. In the initial stages their synodic motion was studied. Astronomers noted the times of their first and last visibility, the duration of their appearance and disappearance, the distance from the Sun at the time of their first and last visibility, the times of their retrograde motion, the distances from the Sun at the times of their becoming retrograde and reretrograde, and so on. Study was also made of their motion in the various zodiacal signs under different velocities called *gatis* (viz. very fast, fast, mean, slow, very slow, retrograde, very retrograde and reretrograde) and along their varying paths called *vīthis*. The synodic motion of a planet was called *grahacāra* and it was elaborately recorded in the astrological works particularly the *saṃhitās*, the earlier works of the Jainas, the earlier *purāṇas*, and the earlier *siddhāntas* such as the *Vaśiṣṭha-siddhānta* and the *Pauliśa-siddhānta*. These records were analysed and in the beginning crude methods or empirical formulae were evolved to get the longitudes of the planets. Later on a systematic theory was established which gave rise to the astronomy of the later *siddhāntas*.

Of the astronomical works written in this period, the *Vaśiṣṭha-siddhānta* is the earliest. *Vaśiṣṭha* and his teachings have been mentioned in the *Yavana-jātaka* of *Sphujidhvaja Yavaneśvara* which was written about 269 A.D. From the summary of the *Vaśiṣṭha-siddhānta* in the *Pañca-siddhāntikā* of Varāhamihira we learn that this work made improvement in the luni-solar astronomy and besides describing the synodic motion of the planets gave empirical formulae for knowing the positions of the planets Jupiter and Saturn. The *Vedāṅga-jyautiṣa* sidereal year of 366 days was replaced by Vaśiṣṭha by the sidereal year of 365.25 days (Neugebauer & Pingree 1971, ii.1). To obtain the Sun's longitude use was made of a Table giving the Sun's motion in the various zodiacal signs (Neugebauer & Pingree 1971, ii.1). The Moon's longitude was obtained in a special way. One anomalistic revolution of the Moon was divided into 248 equal parts called *pada*, each *pada* corresponding to 1/9 of a day. The period of the Moon's one anomalistic revolution was called *gati*, and that of 110 anomalistic revolutions *ghana*. It was assumed that the Moon moved through 111 revolutions -3/4 signs +2 mins. in one *ghana* and 1 rev. (185-1/10) mins. in one *gati*. First the Moon's anomalistic motion since the epoch was obtained in terms of *ghanas, gatis* and *padas*, and then the Moon's motion corresponding to this was obtained and added to the Moon's position at the epoch (Neugebauer & Pingree 1971, ii.2-6). To obtain the Moon's motion for p *padas* in the first half of its anomalistic revolution, the formula used was:

Moon's motion for p *padas* in the first half of its

anomalistic revolution = p degrees+[1094+5 (p-1)]p/63 mins.

And to obtain the Moon's motion for p *padas* in the second half of its anomalistic revolution, the formula used was:

Moon's motion for p *padas* in the second half of its

anomalistic revolution = p degrees+[2414-5 (p-1)]p/63 mins (Neugebauer & Pingree 1971,ii.6).

In the case of Jupiter, starting from the point of zero longitude, its sidereal revolution was divided into 391 equal parts, called *padas*, divided into three unequal segments, the first segment containing 180 *padas*, the second containing the next 195 *padas*, and the third containing the remaining 16 *padas*. When Jupiter was at the end of p *padas* of the first segment, its longitude λ_1 (p) was given by the formula:

$$\lambda_1 \quad (p) = p\ (1456 - p)/24 \text{ mins.};$$

when at the end of q *padas* of the second segment, its longitude λ_2 (q) was given by the formula:

$$\lambda_2 \quad (q) = \lambda_1\ (180) + q\ (1165+q)/24 \text{ mins.};$$

and when at the end of r *padas* of the third segment, its longitude λ_3 (r) was given by the formula:

$$\lambda_3\ (r) = \lambda_2\ (195) + r\ (1486 - r)/24 \text{ mins}$$
(Neugebauer & Pingree 1971,xvii.9-10).

Similarly, in the case of Saturn, starting with the point of zero longitude, its sidereal revolution was divided into 256 equal parts, called *padas*, divided into three segments, the first segment consisting of 30 *padas*, the second consisting of the next 127 *padas* and the third consisting of the remaining 99 *padas*. When Saturn was at the end of p *padas* of the first segment, its longitude $\lambda_1(p)$ was given by the formula:

$$\lambda_1\ (p) = p\ (2416+2p)/27 \text{ mins.};$$

when at the end of q *padas* of the second segment, its longitude λ_2 (q) was given by the formula:

$$\lambda_2\ (q) = \lambda_1\ (30) + q\ (2519 - 2q)/27 \text{ mins.};$$

and when at the end of r *padas* of the third segment, its longitude λ_3 (r) was given by the formula:

$$\lambda_3\ (r) = \lambda_2\ (127) + r\ (2037+2r)/27 \text{ mins}$$
(Neugebauer & Pingree 1971,xvii.16-17).

The above formulae show that at the time of their formulation the longitude of Jupiter's apogee was 165.7 degrees and that of Saturn's apogee 220.8 degrees approximately. In the case of the other three planets no such empirical formulae could be devised and recourse was taken to their motion from one heliacal rising to the next.

A notable feature of the *Vaśiṣṭha-siddhānta* is that it makes use of signs which were not used up to the Vedāṅga period, and reckons the longitudes of the planets from the first point of Aries.

Further progress in astronomy is recorded in the *Pauliśa-siddhānta*. Varāhamihira has described the *Vaśiṣṭha-siddhānta* as inaccurate but the *Pauliśa-siddhānta* as accurate (Neugebauer & Pingree 1971,i.4).

The length of the sidereal year, according to the *Pauliśa-siddhānta*, is 365 days 6 hours 12 seconds (Neugebauer & Pingree 1971, iii.1). This value is better than 365 days 6 hours given by *Vaśiṣṭha*. *Vaśiṣṭha* used approximate rules to get the longitudes of the Sun and Moon. Pauliśa, in the case of the Sun, first obtains the mean longitude and then applies correction for the equation of the centre to get the true longitude. He states a Table giving the equation of the centre for the Sun for the intervals of 30 degrees starting from the point lying 20 degrees behind the point of zero longitude (Neugebauer & Pingree 1971, iii.1-3).

According to *Vaśiṣṭha*, the Moon's motion on the first day of its anomalistic revolution when it is least is 702'; thereafter it increases and reaches the maximum value (Neugebauer & Pingree 1971, iii.4). of 879'. According to the Tables prepared by the followers of Āryabhaṭa the minimum value of the Moon's motion on the first day of its anomalistic revolution is 722' and the maximum daily motion near its perigee[7] is 859', the former being 20' greater and the latter 20' less than the values given by *Vaśiṣṭha*. The values given by *Vaśiṣṭha* are evidently gross. Pauliśa applied two corrections one after the other to the Moon's motion given by *Vaśiṣṭha* but the rules summarized by Varāhamihira have not been understood so far (Neugebauer & Pingree 1971, iii.5.8).

Pauliśa calls the Moon's ascending node by the name "Rāhu's head", and takes 6795 days as the period of its sidereal revolution. The corresponding periods, according to Āryabhaṭa, Ptolemy and modern astronomers, are 6794.7 days, 6796.5 days, 6793 days respectively. The value given by Pauliśa is evidently closer to that of Āryabhaṭa.

The *Paulis'a-siddhānta* deals also with the motion of the planets, the visibility of the Moon and the eclipses. In the treatment of the planetary motion, it gives the distances from the Sun at which the planets rise or set heliacally and become retrograde and reretrograde. The following Table gives the synodic periods of the planets according to Vaśiṣṭha, Paulis'a, Āryabhaṭa, Ptolemy and modern astronomers:

SYNODIC PERIODS IN DAYS

Planet	Vaśiṣṭha	Pauliśa	Āryabhaṭa	Ptolemy	Modern
Mars	779.955	779.978	779.92	779.943	779.936
Mercury	115.879	115.875	115.87	115.879	115.877
Jupiter	398.889	398.885	398.889	398.886	398.884
Venus	583.909	583.906	583.89	584.000	583.921
Saturn	378.1	378.110	378.08	378.093	378.092

Pauliśa's treatment of the visibility of the planets and the eclipses is very approximate.

A notable feature of the *Pauliśa-siddhānta* is the mention of the *viṣuva* and *ṣaḍaśītimukha saṃkrāntis*.

The *Pauliśa-siddhānta* was followed by the *Romaka-siddhānta*. This *siddhānta* bears the impact of the teachings of the Greek astronomers. The day is reckoned from sunset at Yavanapura (Alexandria in Egypt). To obtain the mean positions of the Sun and Moon a luni-solar *yuga* of 2850 years was defined and astronomical parameters were stated for this period (Neugebauer & Pingree 1971, i.15-16). The length of the year used in this work was 365.246 days (Neugebauer & Pingree 1971, viii.1). which is exactly the same as given by the Greek astronomers Hipparchus and Ptolemy. It was really the value of the tropical year but it was used in the *Romaka-siddhānta* as the value of the sidereal year. As the value of the sidereal year it was worse than that given by *Vaśiṣṭha*. The longitude of the Sun's apogee stated in the *Romaka-Siddhānta* (Neugebauer & Pingree 1971, viii.2) was 75°. This was the same as given by Hipparchus when reckoned from the point of zero longitude of Indian astronomy. The period of a sidereal revolution of the Moon's ascending node according to the *Romaka-Siddhānta* was 6796.29 days. This is also almost the same as the value 6796.5 days given by Ptolemy. The maximum equation of the centre for the Sun adopted in the *Romaka-siddhānta* (Neugebauer & Pingree 1971, viii.3,6). was 2°23'23" and that for the Moon 4°56'. The corresponding values given by Ptolemy are 2°23' and 5°1' respectively. Romaka's treatment of the solar eclipse was similar to that found in the later works on Indian astronomy but the rules given are very approximate (Neugebauer & Pingree 1971, viii.). It may be that Varāhamihira himself has condensed them. The *Romaka-siddhānta* did not deal with the planets.

Perfection in astronomy was brought about by Āryabhaṭa who carried out his observations at Kusumapura (modern Patna). He was successful in giving quite accurate astronomical parameters and better methods of calculation. Roger Billard (Billard 1971, pp.81-83) has analysed these parameters and has shown that they were based on observations made around 512 A.D.

Āryabhaṭa wrote two works on astronomy, in one reckoning the day from midnight to midnight and in the other from sunrise to sunrise, in the former dealing with the subject in detail and in the latter briefly and concisely. Both the works proved to be epoch-making and earned a great name for the author. The larger work was popular in northern India and was summarised by Brahmagupta in his *Khaṇḍakhādyaka* which was carried to Arabia and translated into Arabic. This work has been in use by the pañcāṅgamakers in Kashmir till recently. The *Sūrya-siddhānta* which was summarised by Varāhamihira and declared by him as the most accurate work was simply a redaction of the larger work of Āryabhaṭa. The smaller work of Āryabhaṭa called the Āryabhaṭīya was studied in south India from the seventh century to the end of the nineteenth century. This work was also translated into Arabic, by Abu'l Hasan al-Ahwāzī.

Āryabhaṭa's astronomy is based on three fundamental hypotheses viz.

1 That the mean planets revolve in geocentric circular orbits

2 That the true planets move in epicycles or in eccentrics

3 That all planets have equal linear motion in their respective orbits.

Āryabhaṭa's epicyclic theory differs in some respects from that of the Greeks. In Āryabhaṭa's theory there is no use of the hypotenuse-proportion in finding the equation of the centre. Moreover, unlike the epicycles of the Greek astronomers which remain the same in size at all places, Āryabhata's epicycles vary in size from place to place.

The main achievements of Āryabhaṭa are:

1 His astronomical parameters which were well known for yielding accurate results

2 His theory of the rotation of the Earth which was described by him as spherical like the bulb of the kadamba flower

3 The introduction of sines by him

4 His value of $\pi = 3.1416$

5 Fixation of the Sun's greatest declination at 24^0 and the Moon's greatest celestial latitude at $4^0 30'$. These values were adopted by all later Indian astronomers.

6 Integral solution of the indeterminate equation of the first degree viz. $ax+c = b\ y$, a, b and c being constants.

The pattern set by the works of Āryabhaṭa was followed by all later astronomers. The works written by later astronomers differ either in the presentation of the subject matter, or in the astronomical constants which were revised from time to time on the basis of observation, or in the methods of calculation which were improved from time to time. A few new corrections which were not known in the time of Āryabhaṭa were discovered and used by later astronomers. Thus Mañjula (also called Muñjāla) discovered the lunar correction called "evection" and Bhāskara II another lunar correction called "variation".

According to Āryabhaṭa the Sun, the Moon, and the planets were last in conjunction in zero longitude at sunrise at Laṅkā[8] on Friday, February 18, B.C. 3102. This was chosen by him as the epoch of zero longitude to calculate the longitudes of the planets. The period from one such epoch to the next, according to him, is 10,80,000 years. This he has defined as the duration of a quarter *yuga*. Likewise the period of 43,20,000 years is called a yuga. At the beginning and end of this *yuga* the Moon's apogee and ascending node too are supposed to be in conjunction with the Sun, Moon and the planets at the point of zero longitude. The revolution-numbers of the planets stated by Āryabhaṭa are for this *yuga*. The astronomical parameters and the rules stated by Āryabhaṭa are sufficient to solve all problems of Indian astronomy. The main problems dealt with by Āryabhaṭa and other later astronomers are the determination of the elements of the Indian pañcāṅga, calculation and graphical representation of the eclipses of the Sun and Moon, rising and setting of the Moon and the planets, the Moon's phases and the elevation of the Moon's horns and their graphical representation, and the conjunction of the planets and stars.

The ancient Indian astronomers did not possess the telescope. They made their observations with the naked eye using suitable devices for measuring angles. Their astronomy therefore remained confined to the study of the Sun, Moon and the planets.

NOTES

1 According to the Taittirīya-brāhmaṇa 3.1.1, "Jupiter when born was first visible in the nakṣatra Tiṣya (Puṣya)".

2 Other planets are not mentioned by name in the early vedic literature. But Śani (Saturn), Rāhu (Moon's ascending node) and Ketu (Moon's descending node) are mentioned in the Maitrāyaṇī-upaniṣad, 7.6

3 According to Śatapatha-brāhmaṇa, 2.1.2,4, the Great Bear was originally called Ṛkṣa but later the name Saptarṣi was given to it:

4 Of these names the first three occur in Ṛgveda,5.76.3; and sāyam (evening) occurs in Ṛgveda,8.2.20; 10.146.3,40. Kauṭilya (Arthaśāstra, 1.19), Dakṣa and Kātyāyana divided the day and night each into eight parts.

5 Māgha is mentioned in Sāṅkhāyana-brāhmaṇa (= Kauṣītaki- brāhmaṇa) 19.3;Phālguna in Tāṇḍya-brāhmaṇa, 5.9.7-12; and Śrāvaṇa, Māgha and Pauṣa in Ārca-jyautiṣa, 5,6,32 and 34 and Yājuṣa-jyautiṣa, 5,6,and 7; and Mārgaśīrṣa and Śrāvaṇa in Āśvalāyana-gṛhyasūtra, 2.3.1. and 3.5.2. respectively.

6 As regards the origin of the week-days, see Kāne,P.V. (1974) under references.

7 Vide Candravākyāni see Kuppanna Sastri & Sarma (1962) under references. For Candrasāraṇī,see Sūrya-candra-sāraṇī. Ms. No. 1657 of the Akhila Bharatīya Sanskrit Parishad, Lucknow.

8 Laṅkā is the hypothetical place on the equator where the meridian of Ujjain intersects it.

REFERENCES

Aitareya-brāhmaṇa - (1) Tr. Martin Haug, 2 Vols, Bombay, 1863; (2) Ed. Satya-vrata Samasrami with the commentary Vedārthaprakās'a of Sāyanācārya, 4 Vols, Asiatic Society, Calcutta, 1895-1907.

Ārca-jyautiṣa - see Vedāṅga-jyautiṣa

Arthas'āstra of Kauṭilya (1) Tr. English by R. Shamasastry with an introductory note by J.F.Fleet, 4th edition, Mysore, 1951. (2) Ed. and Tr. English with critical explanation by R.P. Kangle, Parts I,II,III, Bombay University, 1960, 1963, 1965.

Atharva-saṃhitā (1) Tr. English by M.Bloomfield as Hymns of the Atharvaveda, Clarendon Press, Oxford, 1897 (2) Ed. Visvabandhu with the commentary of Sayanacarya, Visvesvaranand Vedic Research Institute, 4 vols., Hoshiarpur, 1960-62 (3) Tr. R.T.H. Griffith, 2 vols. Chowkhamba Sanskrit Series Office, Varanasi, 1968.

Billard, Roger (1971). L'Astronomie Indienne, Paris, pp.81.83. 93.

Dvivedi, O. and Sharma, C.L. Atharva-vedīya-Jyautiṣam, ed. vs.

Kane, P.V. (1974). History of Dharmasāstra, **5**, pt.1, Bhandarkar Oriental Research Institute, Poona, pp.677-85; see particularly pp.683-685 where theories about the origin of the seven days in India are given.

Kāṭhaka-saṃhitā. Ed. Schroeder von Leopold, 4 vols, Leipzig, 1909-27.

Kauṣītaki-brāhmaṇa. see 'Sāṅkhāyana-brāhmaṇa

Maitrāyaṇī-saṃhitā. Ed. by Schroeder von Leopold, 2 vols, Leipzig, 1925.

Maitrāyaṇī-upaniṣad - Ed. and Tr. E.B. Cowell, Calcutta, 1870.

Neugebauer, O. & Pingree, D. (1971). Pañcasiddhāntikā of Varāhamihira (d. A.D. 587). Pt.II. Tr. & commentary, Copenhagen.

Ṛgveda - (1) Ed. F.Max Müller, 6 Vols, London, 1857-74; (2) Tr. English by H.H. Wilson, 6 vols, London, 1850; (3) Tr. R.T.H.Griffith, 1896; reprinted in the Chowkhamba Sanskrit Series, Benares, 1963.

Sastry, T.S.K. (1984). Vedāṅga-jyautiṣa of Lagadha, Ed. & Tr. Indian Journal of History of Science, **19** (3 & 4), Supplement.

Sastry, T.S.K. & Sarma, K.V. (1962). Vākyakaraṇa, ascribed to Vararuci (c. A.D. 1300). Critically edited with introduction and appendices, Kuppuswami Sastry Research Institute, Madras.

Śatapatha-brāhmaṇa - (1) Ed. A.Weber with extracts from the commentaries of Sāyana, Harisvāmin and Dvivedaganga, Leipzig, 1924; Second edition, Chowkhamba Sankrit Series No.97, Varanasi, (2) Tr. English by Julius Eggeling, 5 Vols, Sacred Books of the East, **12, 26, 41, 43, 44**, reprinted by Motilal Banarsidas, New Delhi, 1966.

S'āṅkhāyana-brāhmaṇa (=Kauṣitaki-brāhmaṇa) - (1) Ed. ĀnandāS'rama Sanskrit Series, Poona, 1911. (2) Tr. English by A.B. Keith, vide his Ṛgveda Brāhmaṇas, 1920.

Taittirīya-brāhmaṇa - Edited by H.N. Apte with commentary of Sāyaṇācārya, Ānandaśrama Sanskrit Series No.37, 3 vols., Poona, 1898.
Taittirīya-saṃhitā - (1)Ed. Roer and Cowell with the commentary Vedārthaprakāśa of Sāyaṇācārya, 6 vols, Calcutta, 1854-99. (2) Tr. A.B. Keith, Harvard Oriental Series, **18**, **19**,, 1914.
Tāṇḍya-brāhmaṇa(=Pañcaviṃś'a-brāhmaṇa) - Tr. into English by W.Caland, Asiatic Society, Calcutta, 1931.
Tantrasaṃgraha of Nīlakaṇṭha Somayāji (A.D.1444-1545). (1) Ed. with commentary Laghuvivṛti of Śaṅkaravāriar, Trivandrum, 1958, (2) Ed. K.V.Sarma with commentary, Yuktidīpikā and Laghuvivṛti of Śaṅkara, Hoshiarpur, 1977.
Vājasaneyī-saṃhitā (=Śukla Yajurveda-saṃhitā)-Tr. English by R.T.H. Griffith, 1899.
Vedāṅga-jyautiṣa of Lagadha.It has two recensions, Ārca-jyautiṣa and Yājuṣa-jyautiṣa. (1) Ed. with Somākara's commentary by Sudhakara Dvivedi, 1908; (2) Ed. and Tr. English by R. Shamasastry, 1936; (3) Ed. and Tr. English by T.S.K. Sastry, Indian Journal of History of Science, **19**, (3 & 4), Supplement, 1984.
Yājuṣa-jyautiṣa - see Vedāṅga-jyautiṣa

DISCUSSION

L.C.Jain : Could you comment on whether the Jaina Astronomy (*Sūrya Prajñapti, Candra Prajñapti* or *Tiloyapannatti*) was motivated by *Vedāṅga Jyotiṣa* or originated independently?

K.S.Shukla : It was motivated by *Vedāṅga Jyotiṣa*.

2.2 THE ASTERISMS

A.K. Chakravarty
Department of Mathematics, M.R.College, Mahisadal-721 628,
Midnapore, West Bengal, India

A system of 27 asterisms, or *nakṣatras*, plays an important role in Indian astronomy and calendrical science. The present convention is that the ecliptic is divided into 27 equal parts each $13^0 20'$ long commencing from one initial point. These arcs are called *nakṣatras*. Again a bright or prominent star of each division is called *yogatārā* bearing the name of the arc division e.g., *Kṛttikā* as *nakṣatra* means an arc division and as a *yogatārā* it means the star η Tauri of the Pleiades group.

The origin of this asterism system is very old. The names of these stars appear in Vedic litrature. The *Taittirīya, Kāṭhaka, Maitrāyaṇī Saṃhitās* and the 19th book of *Atharva Veda*, each has given a list of 27 *nakṣatras* (the last two have included an additional star, Abhijit (αLyrae), making a total of 28 *nakṣatras*). All these lists always begin with the name *kṛttikās* (the star group Pleiades), and order of names of stars are more or less the same. Some of the stars changed their names in different hands; but they are always either star groups or single stars, and never arc divisions.

In some cases physical descriptions of the stars are given from which they can be identified, and some of these descriptions have astronomical significance which we shall now consider.

The *Kṛttikās* are described as consisting of seven stars (*Taittrīya Saṃhitā* 4.4.5), many stars (*Śatapatha Brāhmaṇa*) and accordingly the *Kṛttikās* have been identified with Pleiades. The *Śatapatha Brāhmaṇa*(2.1.2.2-4) further states that the *Kṛttikās* do not shift from the east. The *Kṛttikās* rise in the east and the seven sages (Ursa Majoris) rise in the north.

Let us take this statement at its face value. If the *Kṛttikās* do not shift from the east, they must be on the equator; we suppose that at the time of observation of this tradition, the central star of Pleiades, η Tauri, had zero declination so that the *Kṛttikās* were seen to rise in the east.

Now, in 1967, latitude of η Tauri = $4^0 2'51''$ N (uneffected by precession), longitude = $59^0 31'53''$. Also, obliquity $\varepsilon = 24^0 1'34''$ in 3000 B.C., and the average rate of precession between 3000 B.C. to 1967 A.D. = 49.7" per year.

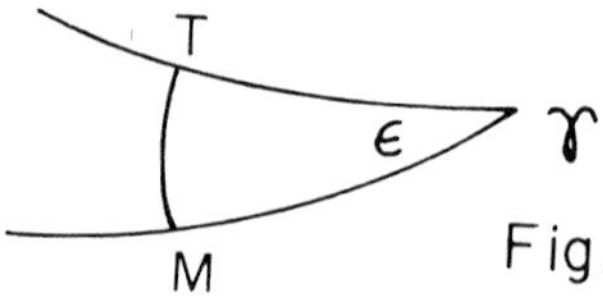

Fig.1

In Fig.1., let γ be vernal equinox, T = position of η Tauri on equator, $\angle \gamma$ MT = 90^0 so that TM = latitude (north) of η Tauri $4^0 2' 51''$ (proper motion of the star and other variations being neglected as they will not affect the result much). From ΔT Mγ,

$$\sin \gamma M = \frac{\tan TM}{\tan \varepsilon} = .15871 \text{ or } \gamma M = 9^0 8'$$

$$\therefore \text{ Total precession} = 59^0 31' 53'' + 9^0 8' = 68^0 40'$$

Hence the epoch comes to about 3000 B.C. at the average rate of 49.7" per year, far before the accepted date of birth of Aryan civilisation in India. Hence we are inclined to believe that the tradition of *Kṛttikās* is very much earlier than the birth of Vedic literature; but anyhow this tradition was handed down to the compilers of *Brāhmaṇas* and *Saṃhitās* who recorded the tradition without any verifications. Our guess-work is that the source from where this tradition reached Aryan India was held in so high esteem that nobody questioned the validity of its truth i.e., it was beyond any question and the tradition was recorded in Vedic literature as a token of respect to the authority of the source.

There is of course a second view. In this opinion, the *Kṛttikās* have been named first in all the lists because of some preferential reason, and this reason is that the *Kṛttikās* at that age were on the equinoctial colure, i.e., vernal equinox was at the *Kṛttikās*. Of course, there is no express or explicit statement to this effect in Vedic literature; a few verses can only be unduly pressed to make out such a meaning. However, proceeding as before and taking η Tauri on the equinoctial colure, we get the epoch at nearly 2400 B.C., which is also earlier than the Vedic period. But in that case, its shift from East Point at rising becomes so much pronounced that it cannot escape any sky watcher's notice. We cannot correlate the statement "*Kṛttikās* rise in the east" and the supposition that the *Kṛttikās* were the vernal equinox. If there was any preferential reason for naming *Kṛttikās* first in the lists, it may be that the *Kṛttikās* rise in the east or that it contain many members or such special properties of it.

Bentley(1823) has given another interpretation of the *Kṛttikā* myth which we shall consider next. In a later section we shall see that there are indeed some evidences, not stated in terms of *Kṛttikās*, which relate to the epoch 2400 B.C.

The astronomy branch of Vedic literature is contained in a small text, the *Vedāṅga Jyautiṣa*. Therein the winter solstice has been placed at β Delphini (Dhaniṣṭhā) which refers to the epoch 1400 B.C. But the period of composition of the text is now supposed to be 600 B.C.

In this text, *nakṣatras* are arc divisions only, and no longer stars or stargroups. The lunar zodiac is divided into 27 equal parts each $13^0 20'$ long commencing from β Delphini (Dhanistha). Lunar positions, *ayanas* etc. are always given in terms fractions of these divisions. *Nakṣatras* are assigned a new meaning arc divisions, and in this scale, winter solstice falls at the beginning of the Dhaniṣṭhā division. The vernal equinox then falls, in this scale of division, at 10^0 of Bharani, i.e., $3^0 20'$ west of the beginning of *Kṛttikā* division.

Bentley's assumption is that at the time of *Śatapatha Brāhmaṇa*, the vernal equinox coincided with this beginning-point of *Kṛttikā* arc division and since then, upto the epoch of *Vedāṅga Jyautiṣa*, there has been a precession of $3^0 20'$, which occurs in 240 years. Accordingly Bentley supposed that the *Śatapatha Brāhmaṇa* was composed around 1700 B.C.

But we are unable to understand it. The main purport of the text, that the *Kṛttikās* rise in the east, does not hold true under this assumption and also, *nakṣatras* are stars and not arc divisions in this text.

We are nowhere told in the Vedic literature the specific purpose for which the asterism system was developed or devised, but there are indications that measures of year, months and fixation of auspicious days were related to many of these stars. Terms like Citrā full-moon-day, or day of Phalgunīs etc. indicate a relationship of a day with one or another star(Taittirīya-saṃhitā,7.4.8) Although the exact purport of the term Citra full moon is not clear, it may mean the day when the full moon disc is visible near the star Citrā in the night sky, or the full moon day in the month of Caitra, yet an attempt of calendarisation of the asterisms is discernible here. It is not known whether the system was devised as a framework for calendar or once the system was ready at hand its services were used for calendrical purposes. But undoubtedly the calendrical scheme was based on this system of asterisms.

Again, we have the following record in the *Maitrī Upaniṣad* (Chap.6) "The Sun makes its southern course from the beginning of Maghā to middle of Sravisthā (i.e. Dhanisthā)". Very clearly, Magā and *Dhaniṣṭhā* here are arc divisions, and not stars. Now, in the *Vedāṅga Jyautiṣa* scale, summer solstice falls at middle of *Aślesā*, and this point is at 4^0 east of the star ε Hydrae. Also from middle of *Aslesa* to beginning of Maghā it is $6^0 40'$. Assuming this $6^0 40'$ to be total precession between the records of the *Vedāṅga Jyautiṣa* and *Maitri Up*., the epoch corresponds to the date 1800 B.C. nearly.

But surely the *Maitrī Up*. did not use the scale of the *Vedāṅga Jyautiṣa* which was formulated some 400 years after it just as an inscription cannot be dated in B.C. years. The text must have used its own zodiac or one formulated before it.

We take this zodiac to have commenced from the star Maghā (Regulus). This initial point cannot be the winter solstice or vernal equinox because then we get a period when human civilisation did not develop. Hence this was the point of summer solstice. The total precession then becomes, between this text and the *Vedāṅga Jyautiṣa*, $13^0 29'$ = Arc betweeen ϵ Hydrae and Regulus ($=17^0 29'$) minus 4^0 The corresponding epoch of the tradition of text becomes about 2400 B.C., a period when the text itself was not compiled.

We now get a highly interesting result. In this case, the vernal equinox was at η Tauri (as it is $89^0 50'$ behind Regulus). Here we have an indirect reference to the fact that the vernal equinox occurs at *Kṛttikās* though no explicit reference to the vernal equinox occuring at *Kṛttikās* is available, as stated earlier.

We thus see that transformation of *nakṣatras* from stars to arc divisions and calendarisation of the system were complete around 2400 B.C. This tradition was also handed down to Aryan India who, without questioning the authority, recorded it in Vedic literature.

Our guess-work is that the practice of determining the nearness of the moon to one or another star by eye-estimation was not considered a reliable method. To avoid any dispute, the stars were replaced by arc divisions, and the eye-estimation method was replaced by a computational method using moon's motion, of course, at its mean rate. We are, however, unable to identify the original source from which Aryan India received these conventions.

We can make a rough estimate of the region from where the above traditions were observationally correct.

The star Svātī (α Bootes) has been described as an outcast (for being far away from other stars) traversing the northern sky(*Mahābhārata*, Śānti-Kalpa 3 - *nityam uttaramārgagam*). The declination of Svātī works out to be $45^0 3'$ in 3000 B.C. (using the transformation formula $\sin\delta = \cos\epsilon \sin\beta + \sin\epsilon \cos\beta \sin\lambda$) and the second to it in declination was α Lyrae, declination being $43^0 38'$. As Svātī had rising and setting, the latitude of the place of observation must be, with some marginal allowances, less than 40^0 It may be noted here that the declinations of these two stars, by 2400 B.C., were $42^0 29'$ and $42^0 4'$ resp. i.e. they were equally northward stars. It is likely that Abhijit was recognised as an additional star around this period, although it was not assigned any arc division.

Again, the record that the seven sages rise in the north seems to bear an inner meaning. By 3000 B.C. α Draconis was the pole star (within tolerable limit of accuracy) and the stars ε and ζ Ursa Majoris were circumpolar almost throughout India, and βU. Majoris had highest co-declination among the seven sages. If we assume that the purport of the text was that all seven sages were then nearly circumpolar or that βU. Majoris was seen just to touch the north point on the horizon at rising, then the latitude of the place should be, after marginal allowances, greater than 30^0; in other words, these traditions were observed from a region within a latitude belt of 30^0 to 40^0.

We explain another tradition in favour of this assumption.

The *Vedāṅga Jyautiṣa* (Yājuṣa, verse 8) has measured the length of the longest day as 14 hours 24 minutes. The unit used in the text is 18 *muhūrtas* where 30 *muhūrtas* = 24 hours. A rough estimation of the place can be made thus:

Half-day = 7 hours 12 minutes = 108^0, declination of sun on longest day = $\varepsilon = 23^0 51' 42''$ (in 1400 B.C.). Hence, from $\cos H = -\tan\phi \tan\delta$, we get $\phi = 35^0$. But this result is unreliable, as effect of refraction has not been considered here. The Indian system is that day begins when the sun is just visible on the horizon. Now, horizontal refraction is 35', and sun's radius is 16'. Hence, at apparent beginning of day, sun's centre will be $90^0 51'$ below the horizon. Now consider the spherical triangle ZPS, where S is centre of sun, Z and P are zenith and pole, and hour angle H is 108^0. Then we get, using cosine rule on $\Delta ZPS, \phi = 33^0 12'$, which we take as 33^0. We neglect dip because any height favourable to development of high civilisation will affect our result very slightly. Thus the tradition of the *Vedāṅga Jyautiṣa* also conforms to our estimation of the latitude belt.

The third and final reformation of asterism system was done in the 6th Century A.D. after which it assumed its present shape where the *nakṣatras* appear both as arc divisions and stars, but as single stars only and not star groups. All astronomial texts of high authority reckon the initial point of these arc divisions from the star ζ Piscium (Brahmagupta, *Bhāskarācārya* and others) or from a point 10' east of it (*Sūrya Siddhānta*). A detailed discussion of this final shape of the system is outside the scope of the present paper.

We have considered so long only such astronomical references from the Vedic literature which are stated in explicit terms and have carefully avoided statements having doubtful or ambiguous meaning, and our finding is that the convention of asterism system was developed before birth of Aryan civilisation in India. This was handed down to Indian Aryans who in turn recorded it in Vedic literature as unquestionable truth. Vedic social life was also adjusted to this convention. We are, however, unable to trace or even to make a guess-work as to how Aryan India received this sytem of asterism; what we know is that the antiquity of this convention has been faithfully preserved in Vedic literature.

REFERENCES

Bentley, John (1823) - Historical view of Hindu Astronomy, Calcutta; reprinted London, 1825.

Mahābhārata - Ed. critically V.S. Sukthankar and others, 22 vols. Bhandarkar Oriental Research Institute, Poona, 1933-59.

Maitrāyaṇī-upaniṣad - Ed. & Tr. E.B. Cowell, Calcutta, 1870.

Śatapatha-bramaṇa - (1) Ed. A. Weber with extracts from the commentaries of Sāyana, Harisvāmin and Dvivedaganga, Leipzig, 1924; Second edition, Chowkhamba Sanskrit Series No.97, Varanasi, 1964; (2) Tr. Julius Eggeling, Sacred Book of the East, **12, 26, 41, 43, 44**; reprinted by Motīlal Banarsidass, New Delhi 1966.

Taittirīya-Saṃhitā - (1) Ed. Roer and Cowell with the commentary Vedārthaprakāśa of Sāyanācārya, 6 vols, Calcutta, 1854-99; (2) Tr. A.B. Keith, Harvard Oriental Series, No. **18, 19,** 1914.

Vedāṅga-jyautiṣa of Lagadha: It has two recensions, Arca-jyautiṣa (1) Ed. with Somākara's commentary by Sudhakara Dvivedi, 1908 (2) Ed. with auto-commentary and English Tr. by R. Shamasastry, 1936; (3) Ed. & Tr. T.S.K. Sastry, Indian Journal of History of Sience, **19** (3 & 4), Supplementary 1984.

DISCUSSION

J.N. Nanda : If the declination angles are 46^0, possible geographical place may be inside India. Any comments.

A.K. Chakravarty : The treatises have not been identified in the pre-Aryan era. The calculated latitude will be the maximum and thus location could be south of the possible 44^0 N.

2.3 VEDIC LITERATURE VIS-A-VIS MATHEMATICAL ASTRONOMY

Ramatosh Sarkar
Birla Planetarium, 96, Chowringhee Road, Calcutta
India.

INTRODUCTION

The present paper restricts to the analysis of some passages from the Vedic literature, viz. *Śatapatha Brāhmaṇa* and *Vedāṅga Jyautiṣa*, from the view point of mathematical astronomy.

The *Śatapatha Brāhmaṇa* in 2.1.2.1 to 2.1.2.5, refers to some ritual in which some fire has to be set up. It recommends *Kṛttikā* as the *nakṣatra* or the lunar asterism under which to set up the fire. For, there are some special features that *Śatapatha-Brāhmaṇa* obviously considers as good points. According to the text, one good point about *Kṛttikā* is that it is 'the most numerous'; secondly, it rises 'in the east'.

Following the time-honoured practice initiated by outstanding Indologists, we can, of course, take *Kṛttikā* to mean Eta Tauri and the stars in its immediate vicinity, that collectively form the 'open cluster' Pleiades. It has six stars visible to the naked eye without much of an effort and some more under good seeing condition. That explains its being described as 'the most numerous'.

But what one may possibly imply by saying, about some stars, that they rise 'in the east'? All stars (that rise) rise in the east. Why then specially mention it about *Kṛttikā*? The conclusion perhaps is inescapable that, in the present instance with reference to *Kṛttikā*, the word 'east' has not been loosely used. Here it implies precisely the 'east point' itself or points very close thereto. In fact, in the present situation, it should have the second, alternative broad implication; because *Kṛttikā* refers to a star-cluster rather than to a single star. But does any of the stars of *Kṛttikā* - say Eta Tauri - really rise at the east point. Now, it does not today. Did it do at any time in the past? Let us examine.

Due to what is known as 'precession of the equinoxes', the celestial longitude of any star keeps on systematicaly increasing. The rate of increase is not uniform. In 2000 A.D. the figure would be $50''.279$ and in 2000 B.C. it was $49''.391$. Let us take the mean of these two last-mentioned values. In other words, for our historical purpose, let the rate be $49''.835$ per year. Now, in 1985, the celestial longitude of Eta Tauri is about $59^0 47' 24''$. But $59^0 47' 24'' \div 49''.835 = 4319$ approximately. Therefore we conclude that about 4319 years back from now, its celestial longitude must have been zero and therefore it would more or less coincide with the vernal equinox. So it could rise at the east point or almost there.

The expressions 'almost there', that has been used here, is not unwarranted. In fact, for Eta Tauri, the celestial latitude is now about $4^{0}2'58''$ and because this co-ordinate can change only very minutely for any star (for Eta, by the way, it is now increasing at the rate of only about $0''.377$ per year), it could have been only very slightly different for Eta Tauri 4319 years back. In other words, in 2334 B.C., Eta Tauri would rise not exactly at the east point but at a point a little to its north - say, 3^{0} to 4^{0} away.

A deviation of 3^{0} - 4^{0} is arguably small - so small as to have been ignored by the ancient people. Another way out is provided by the fact that *Kṛttikā* comprises several stars, of which Eta Tauri happens to be just one. Some other star of the group, more to the south than Eta, might fit the bill exactly and there is nothing in the text to indicate that it was Eta Tauri that was particularly chosen for consideration and not any other star.

The upshot of all this is that circa 2334 B.C., *Kṛttikā* rises "in the east" was a particularly significant statement to make - more significant than at any other earlier time in the history of human civilization.

The *Vedāṅga Jyautiṣa*, in verse 8, indicates that the greatest duration of day-time is 18 *'muhūrta's* (and the shortest is 12). Let us take this statement for scrutiny keeping in mind that, in ancient India, a full day was looked upon as consisting of 30 muhurtas (so that 1 muhurta = .8 hour)

Let us confine ourselves to the northern hemisphere which was the abode of the Vedic people. We know that, for that part of the globe, the higher the latitude, the greater can be the duration of the day-time. Also higher northern declination of the Sun goes with longer day-time. Now, the latitude may be as high as 90^{0} N but the highest solar declination possible is much less. This extreme value of solar declination is subject to some slow periodic change - pendulating from $21^{0}59'$ to $24^{0}36'$. Currently the value is about $23^{0}27'$ but in this historical analysis we shall do well to use the mean value of $23^{0}17'.5$.

Let us examine the issue from different angles by admitting different poosible meanings of 'Sun-rise' and 'Sun-set'.

I <u>WITHOUT CORRECTION FOR REFRACTION</u>

Ordinarily, the Sun is considerd to have risen when it has just wholly come up on the eastern horizon and similarly to have set when it has just wholly gone down below the western horizon.

If the declination of the Sun is δ and the latitude of the place of observation is ϕ (both measured northwards) then (by calculating its 'hour-angle' both at the time of its rise and set) one can write, for its duration of stay above the horizon in terms of hours, the expression $2/15^0 \cos^{-1}(-\tan\phi \tan\delta)$, where the angle denoted by the inverse circular function is assumed to be an angle in degrees.

Putting the stipulated value of δ for the Sun, we then get, according to *Vedāṅga-jyautiṣa*,

$$\frac{2}{15^0}\left[\cos^{-1}(-\tan\phi\tan 23^0 17'.5)\right]^0 = 14.4$$

($\because$ 18 muhūrtas = 14.4 hours)

This yields $\phi^0 = 35^0.66$ nearly.

II WITH CORRECTION FOR REFRACTION

If the effect of refraction is considered, the length of day-time involves apparent Sun-rise and apparent Sun-set. Taking the value of 34' for 'horizontal refraction', the equation in this case changes to

$$\frac{2}{15^0}\left[\cos^{-1}(-\tan\phi\tan 23^0 17'.5)\right]^0 + 2 \times 34/60 \times 1/15 \times \frac{1}{\sqrt{\cos(\phi+\delta)\cos(\phi-\delta)}} = 14.4$$

And if this equation is solved for ϕ, by putting $\delta = 23^0 17'.5$, the computer yields $\phi = 34^0.5$ nearly.

III TAKING THE SUN'S UPPER LIMB FOR THE SUN ITSELF

The Sun, unlike a star of the nocturnal sky, does not appear to be a point of light. It looks like a disc whose angular diameter is about 32' : Sun-rise or Sun-set is not an instantaneous phenomenon.

What is to be done then? There are two distinct procedures. One is to choose any specific part of the Sun. Usually, it is the centre of the solar disc that is chosen. When this chosen part of the Sun rises or sets, then the Sun (as a whole) is considered to have risen or set. The other alternative is to choose what is called the 'upper limb' of the Sun and treat it as representative of the Sun. But in this case, the upper limb does not signify the same part of the Sun throughout: in the eastern sky the point of the Sun first to appear on the horizon is referred to as the upper limb, while in the west the last point to disappear is given that appellation. It is by far the more popular stand: to people in general, as soon as a bit of the Sun is seen, it is day begun; as long as the last bit of it is seen, the day is still on.

If we adopt the first procedure, which in fact is the more rational of the two, we are de facto treating the Sun as a star of the night-sky and therefore the equation and the solution of II hold in toto.

If, on the other hand, we assume that the observers of *Vedāṅga-jyautiṣa* took the usual or popular view, then to them the day-time got unduly lengthened. If we want to correct for this error, our equation becomes

$$\frac{2}{15^0}\left[\cos^{-1}(-\tan\phi\tan 23^0 17'.5)\right]^0 + 2 \times (34+16)/15 \times 1/60 \times \frac{1}{\sqrt{\cos(\phi+\delta)\cos(\phi-\delta)}} = 14.4$$

Replacing δ by $23^0 17'.5$ and then solving this equation for ϕ by means of computers, one gets $\phi^0 = 34^0$ nearly.

CONCLUSION

The results stemming from the aforesaid analysis and calculations are somewhat astonishing. For, modern scholars, by and large, do not associate Vedic literature with the third millennium B.C., which is considered rather to be the flourishing period of the Indus valley civilization. Also, Vedic people are geographically associated with lower northern latitudes.

A plausible conclusion is that the relevant astronomical observations were made much earlier and their narration in literary form took place later. The Aryans probably observed while they were still in the process of emigration, or they carried the information from their earlier settlements, or else they got it from the Harappan people.

REFERENCES

Indian Astronomical Ephemeris (1985). Pub. India Meteorological Department, Govt. of India.

Śatapatha-brāhmaṇa - Ed. Julius Eggeling, Sacred Book of the East, 1882; reprinted Delhi, 1972.

Vedāṅga-jyautiṣa - Ed. Sitesh Chandra Bhattacharya, Sanskrit College, Calcutta, 1974.

2.4 THE CHARACTERISTICS OF ANCIENT CHINA'S ASTRONOMY

Xi Zezong
Institute for History of Natural Science
Academia Sinica, Beijing
China

What are the characteristics of ancient China's astronomy? Many scholars have discussed the problem. In 1939 Herbert Chatley summed up fifteen points. Joseph Needham(1959) in his great work *Science and civilisation in China* concentrated it into seven points.

1 the elaboration of a polar and equatorial system strikingly different from that of the Hellenistic peoples;

2 the early conception of an infinite universe, with the stars as bodies floating in empty space;

3 the development of quantitative positional astronomy and star catalogues two centuries before any other civilisation of which comparable works have come down to us;

4 the use in these catalogues of equatorial coordinates, and a faithfulness to them extending over two millennia;

5 the elaboration, in steadily increasing complexity, of astronomical instruments, culminating in the 13th century invention of the equatorial mounting, as an 'adapted torquetum' or 'dissected' armillary sphere;

6 the invention of the clock drive for that forerunner of the telescope, the sighting tube, and a number of ingenious mechanical devices ancillary to astronomical instruments;

7 the maintenance, for longer continuous periods than any other civilisation, of accurate records of celestial phenomena, such as eclipses, comets, Sunspots, etc.

Needham also pointed out that the most obvious absences from such a list are just those elements in which occidental astronomy was strongest: the Greek geometrical formulations of the motions of the celestial bodies, the Arabic use of geometry in stereographic projections, and the physical astronomy of the Renaissance.

Liu Jinyi (1984) also put forward ten points which are similar to those of Herbert Chatley. Their viewpoints, I think, describe its contributions rather than characteristics. As regards the fundamental characteristics, they were not clarified. Zhu Kezhen (1951) considered that there were two fundamental characteristics, i.e. practical application and protracted nature; but he did not mention it in detail. I suppose the former is more important than the latter and shall discuss it here.

At the present time astronomy is a pure science and belongs to fundamental sciences, but the early period of a science in its development is always different from the late period. Thomas S. Kuhn (1979) pointed out that "Early in the development of a new field, social needs and values are a major determinant of the problems on which its practitioners concentrate. Also during this period, the concepts they deploy in solving problems are extensively conditioned by contemporary common sense, by a prevailing philosophical tradition, or by the most prestigious contemporary sciences." In China, astronomy originated in the need of agriculture and astrology, and due to the influence of Chinese social conditions and traditional culture, developed in a way quite different from that of Greek astronomy. From the school of Pythagoras (C 582-500 B.C.), Greek astronomy intended to set up a model of the universe, while Plato (427-347 B.C.) further saw that any philosophy with a claim to generality must include a theory as to the nature of our universe. But in so doing, just as S.A.Mason (1953) pointed out, he did not wish to stimulate the observation of the heavens; on the contrary, he desired only to make astronomy a branch of mathematics. This ideological line determined the rationalism of European astronomy; though later astronomers took to the observation of the heavens to obtain data for calculation, test and improvement of their models of the universe.

On the contrary, natural philosophy in China did not develop to the full and occupied no distinguished position (cf. Ye Xiaoqing, 1984). Chinese sages only wanted astronomers "to observe the heavens so as to investigate the change of human affairs on the earth" (The Classic of Changes, *Yi Jing*) as well as "to observe the Sun, Moon and stars in order to issue the official calendar" (the Yao Canon of the *Shu Jing* Historical Classic). This ideological line determined the pragmatism of Chinese astronomy.

On the other hand, since the calendar reform directed by Julius Caesar in the middle of the first century B.C. the calendar used in Europe is a solar one, and which developed into the Gregorian calendar, and only requires the accuracy of tropical year in order to coordinate the relation between the days and the months, regardless of the motion of the Moon and the planets, so the calendar making occupies a very small position in the western astronomy. In contrast with that, as far back as the 14th century B.C. China had an embryonic form of a lunisolar calendar, and from the second century B.C. Chinese calendar contained the fundamental contents of modern astronomical almanac, including the calculation and observation of the positions of the Sun, Moon, planets and stars and of solar and lunar eclipses, so Chinese astronomy developed by the way of calendar making and had a character of applied science.

The differences of Chinese lunisolar calendar from that of Babylon and Greece are (1) to take "shuo" (the moment when the Sun and the Moon are at the same longitude) as the beginning of a month; (2) to think of the winter solstice point as starting one for measurement of solar apparent position as well as star positions; (3) to fix the moment of the Sun at the winter solstice point in the eleventh month, and to take this moment as the beginning of a year, and from which to divide a tropical year into 24 periods (12 "jie" and 12 "qi", solar terms). Of these 20 names are connected with season, air temperature or precipitation, such as "lichun" (the beginning of spring), "dashu" (great heat), "xiaoxue" (slight snow). They directly show the change of seasons, and make agricultural things very convenient. The system so far helps farmers to know what kind of weather to expect in each period.

The 24 solar terms are directly determined by the apparent position of the Sun and belong to the category of the solar calendar. For coordinating the relation between them and the synodic months it is necessary to arrange the intercalary month ("run"), therefore qi (12 solar terms), shuo and run make up the three elements of Chinese calendar, around which Chinese astronomy developed. The solar terms can be measured by Gnomon, because the solar shadow of gnomon is longest at winter solstice and shortest at summer solstice. At the moment when the Sun and the Moon have the same longitude, the Moon cannot be seen. Only when a solar eclipse takes place,it can be proved that the Sun and the Moon have same longitude and same latitude, so observation and calculation of eclipses became an inseparable part of Chinese calendar-making . On the one hand, for raising the accuracy of prediction of the solar terms, the beginning of a month, the intercalary month and eclipses, it was necessary to improve the calculation method; on the other hand, for raising the precision of their observation, new instruments had to be made and new observational methods had to be invented. And both sides complement each other. According to study by Chen Meidong (1983), the evolution of Chinese calendar can be shown in Table 1.

TABLE 1

Error of / Time	qi	shuo	eclipse	planetary position
B.C.206-A.D.220	3-2 days	1 day	1 day	8^0
220-589	2-1/5 day		15-4ke*	$8-4^0$
581-1127	20-10ke		4-2ke	$4-2^0$
1127-1368	10-1ke		2-0.5ke	$2-0.5^0$

* 1ke = 14^m4

Calendar making in China not only served agricultural production, but also formed a part of the superstructure. To promulgate the calendar was a symbol of dominion and only the court should have it in hand. To use the calendar promulgated by an emperor meant recognition of his political power. In the calendrical Chapter of *Shiji* (Historical Records) Sima Qian said: "When a new dynasty was established, the emperor must change his surname, alter the calendar, transform the colour of ceremonial dress and calculate the position of vigour in order to undertake the mandate of heaven". After the emperors You and Li, the Zhou dynasty declined and the emperor did not promulgate the shuo, so the calendar of the Lu dukedom, which used the shuo promulgated by the Zhou emperor, was not corrected either. In the sixth year of Wengong of the Lu dukedom (621 B.C.) the Duke did not promulgate the shuo of the intercalary month. About this matter the classics *Zuozhuan* (Master Zuoqiu's Enlargement of the *Spring and Autumn Annals*) wrote critical remarks as follows:

"Not to inaugurate solemnly the first day of the intercalary month was an infringement of the proper rule. The intercalary month is intended to adjust the seasons. The observance of the seasons is necessary for the performance of the labours of the year. It is those labours by which provision is made for the necessities of life. Herein then lies caring for the lives of the people. Not to inaugurate properly the intercalary month was to set aside the regulation of the seasons; --- what government of the people could there be in such a case?" (English translation by Legge 1872).

By the end of the period of Spring and Autumn, Zigong, a student of Confucius, wanted to cancel the system of offering a sheep for promulgating the shuo. Confucius opposed it and said "You love the sheep but I love the rite". Right up to the seventeenth century, when the Qing Court appointed Adam Schall (a German Jesuit, 1591-1666) to calculate the official ephemeris, because of the five Chinese characters "Yi Xi Yang Xin Fa" (Based on the New Western Method) printed on the front cover of the ephemeris, Yang Guangxian, a scholar from Anhui province, accused him of usurping state power and of stealing secret information under the cover of compiling the calendar, thus causing consternation in the Qing Court. Consequently, on the 1st day of the 4th month of 1655, the ministries of Rites and Punishments drew up a proposal, according to which Adam Schall should be put to death by dismemberment. On the next day while the Regents were holding a meeting to ratify the proposal, they had to flee in alarm from a sudden earthquake. Thereafter earth quakes continued from time to time and a comet appeared in the sky. According to traditional Chinese astrology, the Qing court regarded these phenomena as manifestations of the anger and discontent of Heaven, and offenders must have their penalties reduced. Hence Adam Schall and his assistant Ferdinand Verbiest (1623-1688) were released from prison and then took charge the Royal Bureau of Astronomy again (cf. Xi Zezong 1982).

Similar to the case of Babylonia, Chinese astrology belongs to the judicial or portent system (cf. Nakayama S. 1966), which by the observation of celestial phenomena (especialy abnormal) divines important events, such as victory or defeat in a war, the rise and fall of a nation, success or failure of the year's crop, and the actions of emperor, empress, concubines, princes, feudal lords and court officials. Here astronomy played a check on the ruler, and astronomers were regarded as interpreters of celestial signs (cf. Eberhard D. 1957). For the former case, we can take an example from the astronomical Chapter of *Shiji*: "When Mercury appears in company with Venus to the east, and when they are both red and shoot forth rays, then foreign kingdoms will be vanquished and the soldiers of China will be victorious. When they are to the west, it is favourable to a foreign country". For the latter, we can take another example from the "Five Elements" Chapter of *Hanshu* (History of the Han Dynasty): On the First day (Wushen) of the 12th month of the third year of the Jianshi era of the reign of Emperor Cheng (January 5, 29 B.C.) a solar eclipse took place in the sky and in that night an earthquake occurred in the Weiyang palace. Astrologer Gu Yong reported to the Emperor: "Solar eclipse at 9^0 of Wunu (the 10th lunar lodge) means that something will be wrong with the empress, while an earthquake within the screen wall lays the blame on a noble concubine. Now both of them take place at the same time, portending that Yin will make attack upon Yang. I think, the empress and the concubine will together do the prince harm." When the Emperor asked another astrologer Du Qin, Du also said: "The solar eclipse happened at Wei (13^h-15^h) on Wu Shen day. Wu and Wei represent earth (one of the Five Elements) which corresponds to the central region. Combining it with the fact that the earthquake occurred within the palace, I suppose the close concubines will do harm each other so as to contend for the love of the Emperor, and when affairs go wrong on Earth, abnormal phenomena appear in the Heaven. If the emperor undertakes moral conduct, the disaster can be eliminated by itself. If he neglects the warning of the Heaven and does not care it, the disaster will come".

The sayings of Gu and Du represent the difference between Chinese and Babylonian astrology in that the theoretical foundation of Chinese astrology is the Yin-yang theory, the Five Elements theory and the heaven mandate theory. The Yin-yang theory explains all the phenomena in the universe in terms of a fundamental dichotomy which corresponds to that between heaven and earth, male and female, and so on. The five elements theory was used to systematize the relations of things by placing them in the constellation of natural agents - wood, fire, earth, metal and water. The heaven mandate theory considers that the monarch is a man of transcendent virtue, whose title to the throne is bestowed by heaven, in other words, he is the agent of the natural order and he rules under its auspices. If, then, his conduct is contrary to the natural order, he is no longer qualified for the throne. In this respect, the royal astronomers were emperor's advisors, and celestial phenomena were matters of great concern to the throne, implying grave political consequences. For example, please read the proclamation of the Emperor Wen of the Han dynasty in 178 B.C. for a solar eclipse: "We have heard that when heaven gave birth to the common people, it established princes for them to take care of and govern them. When the lord of men is not virtuous and his dispositions in his government are not equable, Heaven then informs him by a calamitous visitation, in order to forewarn him that he is not governing rightly. Now on the last day of the eleventh month there was an eclipse of the Sun—a reproach visible in the sky—what visitation could be greater?...Below Us, we have not been able to govern well and nurture the multitude of beings; above Us, we have thereby affected the brilliance of the three luminaries (i.e., the Sun, the Moon, and the stars). This lack of virtue has been great indeed. Wherever this order arrives, let all think what are Our faults and errors together with the inadequacies of our knowledge and discernment. We beg that you will inform and tell Us of it and also present to Us those capable and good persons who are foursquare and upright and are able to speak frankly and unflinchingly admonish Us, so as to correct Our inadequacies. Let everyone be therefore diligent in his office and duties. Take care to lessen the amount of forced service and expense in order to benefit the people." (English translation by Dubs H.H. 1938). This is the beginning of a new system of selecting talented persons for government, which lasted many centuries.

Since celestial phenomena were so important to the state affairs, astronomical work was of course given much attention and became a part of government work. From about 2000 years B.C. an observatory was established. During the regime of Qin Shihuang (259-210 B.C.), the first emperor of China, there were over 300 persons engaged in astronomical observations in court. According to *Jiu Tang shu* (The Old History of Tang Dynasty), at that time (618-907 A.D.) as a bureau, the royal observatory worked under the direction of the Department for the Imperial Archives and Library and consisted of the following 4 parts:

1 calendar making: 63 persons
2 astronomical observations: 147 persons
3 time-keeping (managers of clepsydras): 90 persons
4 time-service(reporting time by bell and drum):200 persons

It is a characteristic of ancient Chinese astronomy that there were so many astronomical workers in a government and their heads had positions of such high rank. This characteristic was at first sight noted by the Jesuit Matteo Ricci of Italy (1552-1610), who used it to do missionary work. He never ceases saying that "astrology" was generally practised by the Chinese society of his time, and it would have been an error not to see in this somewhat inappropriate term all the social importance and philosophical elevation with which it was clothed in the Far East (Cf. Bernard,H., 1935). Matteo Ricci wrote on May 12, 1605 to a correspondent in Europe:

"I address to Your Reverence urgent prayers for a thing which I have for long requested and to which I have never received any reply: it is to send from Europe a Father or even a Brother who is a good astronomer. In China the king maintains, I believe, more than 200 people at great expense to calculate the ephemerides each year. If this astronomer were to come to China, after we had translated our table into Chinese, we would undertake the task of correcting the calendar, and, thanks to that, our reputation would go on increasing, our entry into China would be facilitated, our sojourn there would be more assured and we would enjoy great liberty". At the invitation of Matteo Ricci, missionaries who had a good command of astronomy arrived in China in 1620 and European astronomy began to be widely introduced into China.

In Europe national observatories were built only from the end of the seventeenth century. In the Islamic world no observatory existed for more than 30 years, and it always declined with the death of a king. Only in China did the royal observatory last thousands of years, in spite of the changes of dynasties.

Not only the royal observatory but also the astronomical records lasted thousands of years. The Chinese term "tian wen" is now used simply to mean "astronomy", but this is a decided shift in connotation; in classical writings it is ordinarily used in the sense of portent astrology. The major part of the Tianwenzhi (astronomical chapter) in the official histories is devoted to the observational records of celestial abnormal phenomena and their connection with political events. As a result, the 24 Histories preserve voluminous and detailed collection of such observations - a collection which has so far attracted the attention of scholars both in China and abroad, and some remarkable results have been achieved when it is related to modern astronomical problems such as the remnants of supernovae, solar activity and so on (cf. Xi Zezong, 1983).

Apart from the astronomical chapters in the 24 Histories, there are Lizhi (calendrical chapters) in which there is described how to compute the motion of the Sun, Moon and planets, how to predict eclipses and how to observe these phenomena and stellar positions.

In summary we can say, ancient Chinese astronomy mainly comprised two parts: calendar making and celestial phenomena observation, instrument making was in the service of these two tasks. These tasks were considered a part of political affairs, the royal observatory was one of government departments, and the heads of astronomical profession were royal advisors, men of high rank and position. They were not interested in pure science for science's sake, and they did not spend enough time in developing abstract laws. So the development of astronomy in China was closely connected with the feudal society and could not be expected to transform into modern astronomy.

REFERENCES

Bernard, H. (1935). Matteo Ricci's Scientific Contribution to China p.54, Beijing.

Chatley, H. (1939). Ancient Chinese Astronomy, Occasional Notes of R.A.S., **5**, pp.65-74.

Chen Meidong, (1983). Observation Practices and Evolution of the Ancient Chinese Calendar, Lishi Yanjiu, **4**, 85-87.

Dubs, H.H (1983). The History of the Former Han Dynasty. **1**, p.240-1, Baltimore, Waverly Press.
Eberhard,D. (1957). The Political Function of Astronomy and Astronomers in Han China. In Chinese Thought and Institutions edited by John K. Fairbank, pp.33-77, Chicago.
Kuhn,T.S. (1979). The Essential Tension, p.118, Chicago.
Legge,J. (1872).tr. The Chinese Classics translated **5**, part 1, p.245, London.
Liu Jinyi et al., (1984). Astronomy and its History (in chinese) pp.34-5, Beijing.
Mason,S.A. (1953). A History of the Sciences, pp.23-24, London.
Nakayama,S. (1966). Characteristics of Chinese Astrology, Isis, **57(4)** 442-454.
Needham,J. (1959). Science and civilisation in China, **3**, p.458, Cambridge.
Ricci,M. (1913). Opere storiche, **2**, pp.284-5, Macerata.
Xi Zezong, (1982). Verbiest's Contribution to Chinese Science; Proceeding of the First International Conference on the History of Chinese Science, Leuven, Belgium. (in press).
Xi Zezong, (1983). The Application of Historical Records to Astrophysical Problems, Proceedings of Academia-Sinica-Max Plank Society Workshop on High Energy Astrophysics, Beijing, pp.158-169.
Ye Xiaoqing, (1984). On the Position of Science and Technology in Traditional Chinese Philosophy; Proceedings of the Third International Conference on the History of Chinese Science, Beijing. (in press).
Zhu Kezhen, (1951) Ancient China's Great Contributions to Astronomy, Kexue tongboa, **2(3)**, 215-219.

DISCUSSION

S.S.Lishk : Were angular distance of heavenly bodies measured in terms of earth distances 'Li' ?

Xi Zezong : No.

L.C.Jain : In Chinese calendrical calculations two words, Phing Chhao (floating difference) and Ting Chhao (fixed difference) were in use. Could you kindly comment on whether any other Asiatic nation made use of these words during the contemporary period or earlier periods ?

Xi Zezong : These were the developments of later period in China.

S.Nakayama : Does ten days difference make really a success or failure of crops ?

Xi Zezong : Difference of one unit value in astronomical system will lead it into crisis.

S.D.Sharma : Yuga of 60 year cycle has been used both in Indian and Chinese tradition. Could you kindly explain how the same was used in Chinese Calendar ?

Xi Zezong : I think that the two 60 year cycles have no common origin.

2.5 ANCIENT CHINESE AURORAL RECORDS: INTERPRETATION PROBLEMS AND METHODS

Michel Teboul
C.N.R.S. and Tokyo Astronomical Observatory,
The University of Tokyo,
Japan.

Abstract: In recent years a number of auroral catalogues have been compiled from Oriental records, but the auroral identifications given vary greatly from one catalogue to another. This is due to the fact that, up to now, no reliable criteria have been found which could help us discriminate between auroral and non-auroral ancient records. In this paper two new criteria are propounded and their usefulness illustrated by the solution of a few currently much debated questions.

I INTRODUCTION: CURRENT ORIENTAL ARCHAEO-AURORAL STUDIES.

In recent years great interest was aroused in the study of Chinese, Korean and Japanese historical records on the Aurora Borealis in order to obtain new data on the secular variation of the geomagnetic dipole field. It was hoped that by comparing these records with the many catalogues that have been compiled of the historical observations of the Aurora Borealis in the Occident, such as those by Hermann Fritz (1881) and Frantisek Link (1962,1964), one could find examples of simultaneous appearances of the Aurora Borealis in the Orient and the Occident, and indeed a few such cases have already been detected and discussed(Keimatsu et al.1968).

This method was first applied by *Mitsuo Keimatsu* [a] and *Naoshi Fukushima* [b], and after the publication in 1968 of their preliminary results, *Keimatsu*(1970) started to compile a list of all sunspots and auroral records he could find in Oriental sources written in classical Chinese . His work soon attracted the interest of Chinese scientists and historians of science who, finding that they could not agree with many of *Keimatsu's* auroral identifications, started compiling their own lists of auroral records for the Orient. Up to now three such systematic catalogues have been compiled(see Reference: Articles in Chinese), all at variance with one another and with *Keimatsu's* catalogue.

It must be noted here that the exploration of Oriental auroral records is in fact as old as Occidental sinology. Just after the publication in France of the first book ever devoted entirely to the Aurora Borealis and its history, the famous *Traité Physique et Historique de l'Aurore Boréale* (Paris,1733), its author, Jean Jacques Dortous de Mairan, sent a copy to Antoine Gaubil (1689-1759)[1] who seems to have received it around the beginning of August 1736(Gaubil 1970,p.473). Mairan must have asked Gaubil to look in Chinese records for observations of the Aurora Borealis and consequently Gaubil was to give two auroral identifications in his *Histoire de l'Astronomie chinoise, depuis le commencement de la monarchie chinoise jusquā 1'an 206 avant J.-C.* which he sent to France in 1754, one copy being sent to Mairan.[2] After Gaubil, another French sinologue, Edouard Biot (1803-1850), who was very much influenced by Gaubil's works, took up the matter again in 1844 and extracted from Chinese sources over 40 records which he identified with Aurora Borealis and of which he published only two(Needham 1954, p.482), but no one in the Occident seems to have pushed further in this way after him, perhaps because nowadays the Aurora Borealis is a very rare occurrence in China(Teboul 1985).

II TWO NEW AURORALNESS CRITERIA.

It would seem that neither Gaubil nor Biot bothered much about giving reasons for their auroral identifications[3], *Keimatsu* being the first to be fully aware of the great difficulty of the precise identification of the phenomena described in the original sources with an Aurora Borealis. This he tried to overcome by giving each record a reliability coefficient, which was not a very happy solution as in most cases it still left his reader facing the same problem as at the outset, namely, which record should be treated as an auroral one, and which should be discarded. In fact, Chinese scientists disagreed with many of *Keimatsu's* auroral identifications, the relevant records being rather, they argued, records of meteor showers, comets or even plain meteorological phenomena . The reasons they invoke all deal with some aspect or another of the physics of the upper atmosphere, but, as a matter of fact, it is because of this wealth of theoretical considerations that the three catalogues so far compiled in mainland China, all mostly based on the same primary sources[4] differ greatly in their results, each compiling team having chosen, from the beginning, its own set of assumptions on what it should treat as an Aurora Borealis and what it should rather classify into another category of celestial phenomena. By so doing they forgot that the records we are dealing with were, after all, written a long time ago, in a language vastly different from even modern Chinese, by people whose views on Nature were perhaps not the same as ours and whose recording aims were much more astrological than astronomical. Besides, the Aurora Borealis being a phenomenon which can display an infinite variety of shapes, colours, structures and movements, we must except to come up with descriptions couched in a language too strange to be immediately apprehended as referring in fact to some kind of auroral event. It would thus seem highly advisable, as far as auroralness criteria are concerned, to shift our emphasis from pure physics to: first, a re-examination of the records from the point of view of traditional Chinese philology, in order to get some fresh insights into their real meanings, that is to say the meaning they had for the people who wrote them down as observational records, and second, a comparison with modern auroral descriptions[5], in order to confirm that our tentative philological reconstruction is compatible with the mechanisms of Aurora Borealis[6]. I would like to present briefly here one case study and a few implications of these two criteria, in order to illustrate this entirely new way of tackling suspected ancient Chinese auroral records.

III THE AURORAL RECORDS IN THE BOOK OF CHANGES

The date of the *Book of Changes*[c] is still a highly disputed matter because it is in fact an amalgam of elements whose different origins are distributed over a great span of time and space. It has already been shown by Chinese scholars that some of these elements refer to natural phenomena, but it seems to have escaped previous researchers that the text of the second Hexagram, *Kūn*[d], seems, at least in part, to describe some auroral display of a band travelling upwards from the horizon until it develops into a drapery with ray-structure. Though the Chinese original, being extremely concise, is difficult to interpret, we may try to give for it the following rendering[7](Sung 1973,p.17).

A First line: (After) treading on frost we arrive(d) on strong ice. This gives the setting, which looks very much like a description of some travel towards the land of "eternal winter".

B Second line: In front, aside, (all was) vastness. The travel into the frozen vastness goes on.

C Third line: (That which) contains brightness (and to which) can be put questions (shows). This is the first hint of the Aurora Borealis. Incidentally, we have here the motive of the travel up north referred to in A. and B. :the search for the oracle Aurora Borealis was considered to be.

D Fourth line: (It is a) fastened-up bag. It does not drum, it does not expand[8]. Attentive looking shows that this brightness emanates from a "close bag" of clouds, low on the horizon. No sound comes out of it, and it does not move.

E Fifth line: A yellow lower garment (shows). The homogeneous arc located near the horizon shows a yellow-green lower border.

F Sixth line: Dragons are fighting in sight. Their blood (is like) dark-purple and yellow rays. The arc has risen towards the zenith and is developing now variegated rays.

We can now fulfil the second part of the proof by producing the following extracts from a well known book(1885,p.203) by Sophus Tromholt (1851-1896):

1.b. "..., and a solemn silence reigns over the endless icefields which surround us on all sides."

2.a. "Even the snow seems to have become ice,..." Which parallel very well with the text of the Second and First Lines, respectively.

3.d. "In some part of the horizon or another lies a close cloud."
This sentence is a striking analogy with the Chinese expression found in the Fourth Line, "a fastened-up bag".

4.c. "Its upper edge is illuminated,...."
This explains very well the text of the Third Line.

5.e. "..., the band slowly but steadily alters its position and form. The width is considerable, and the intense whitish-green light stands out in magnificent relief on the dark background."
This is but an expanded statement for the text of the Fifth Line.

6.f. "The band now falls into manifold curls,... Waves of light course constantly through the entire length of the band with an undulating motion, now they run from right to left, now from left to right; they seem apparently to cross each other as they appear on the nearest or furthest side of a curl."

This should not be so difficult to interpret in terms of dragons, fighting one with another. Such a fight must be associated with the shedding of blood, hence the end of the text of the Sixth Line which would appear to describe a drapery with ray-structure.

This kind of analysis has for immediate consequence the fact that the text of the first Hexagram, *Qián*[e], describes also an auroral display, the only difference being that the emphasis there is put on the word "dragon" which, in the text of the second Hexagram, appears only once.

That the word "dragon" referred in fact to Aurora Borealis in the *Book of Changes* is not to surprise us, since in Europe also many auroral displays were, quite rightly if we think of the motions of a folded band, described in terms of dragons(Eather 1980, p.44), and is fully corroborated by many texts up to the end of the Posterior *Hàn* dynasty(Teboul 1985).

IV RECORDS IN TERMS OF "STARS FALLING LIKE RAIN".[f]

The first record on which opens *Keimatsu's* catalogue is a record of "Stars Falling Like Rain" which Gaubil, sometimes between 1735 and 1754, had already identified as some kind of Aurora Borealis[2]. Many such records can be found in Chinese sources but Chinese scientists have decided, of late, that it would be safer not to interpret them as auroral descriptions but rather as records of meteor showers. *Keimatsu* on his part based his auroral identification on the fact that in the second Official Dynastic History of China there are three different records of what he took to be the same phenomenon observed on the 27th of March, 15 BC (Julian style) (Keimatsu 1970, record no 16, Keimatsu gives the date as March 23), one of which is unmistakably an auroral record, while the other two are in terms of "Stars Falling Like Rain". Making then the explicit assumption that all three records were extracts from one and the same observation report which had later been arbitrarily distributed between three different parts of this second Official Dynastic History of China (Keimatsu 1970, pt.1,p.3), he inferred: first, that the two extracts in terms of "Stars Falling Like Rain" were simply to be equated with the third one, and second, that the majority of Chinese historical records using this same technical expression was auroral in nature.

Needless to say, *Keimatsu's* assumption must have been based on the fact that there is a very low probability for an aurora and a meteor shower (or shooting stars) to be both observed and reported on one and the same night. It is true that this is a rare occurrence, but it is not unknown and we can quote an observation report from Camille Flammarion (1842-1925) describing just such a coincidence during the auroral display which was seen over Paris on the 13th May, 1869 (Flammarion 1888, p.766).

> "Quelques étoiles filantes ont signalé cette période[10]. Un bolide est parti du voisinage du zénith à onze heures trente-cinq minutes, pour s'éteindre en arrivant à la hauteur de la Grande-Ourse. Un autre a semblé tomber de Véga à onze heures quarante-cinq minutes."

Another, sinological, argument may be given against *Keimatsu's* assumption: it can be shown that the Chinese historical works which are our main sources of auroral records, such as the *Monographies on Astronomy*[g], or the *Monographies on the Five Elements*[h] of the successive Official Dynastic Histories of China, are themselves very strictly structured in terms of the astrological theory prevalent at the time when they were compiled, each sub-section of these Monographies dealing in fact with one and the same type of astrological event. It is therefore impossible for the same *astronomical* event to be listed under two or more different *astrological* headings[11], since a given astronomical event could have one, and only one, *type* of astrologically associated meaning[12].

Though it is thus quite clear that on the 27th March, 15 BC, there had been two different celestial phenomena duly reported by the Imperial Observatory and later recorded in the appropriate parts of the second Official Dynastic History of China, which contradicts *Keimatsu's* basic assumption, it is nevertheless quite probable that his conclusion, namely that (if not all, at least some of) the historical records using the technical expression "Stars Falling Like Rain" must be records of auroral displays, is correct.

To prove this we must first notice that the Chinese character *xīng* 星 which, even in the context of Aurora Borealis is nearly always translated as "Star", may have in fact quite different meanings, the best translation being probably the word "Meteor" as it was understood before it took on its modern technical meaning of a "shooting star"(Eather 1980, p.61)[13]. The Chinese expression under consideration here should thus read (in first approximation) "*Meteors* Falling Like Rain" which we may accept as referring, under appropriate conditions, to some kind of auroral display of the type seen over Iceland on the 20th

August, 1886, and described by the French explorer Noël Nougaret(Flammarion 1888,p.764).

> *"L'aurore est alors dans tout son éclat; du ciel se détache (sic) de longues franges qui descendent mollement et que l'observateur croit pouvoir saisir avec les doigts."*

Where the original French "de longues franges qui descendent mollement (du ciel)" fits in perfectly with Peter Van Musschenbrock's (1692-1761) general definition of the *meteor*(Eather 1980,p.61). Indeed, the same kind of metaphor flows naturally from the pen of Flammarion when, describing the light of the aurora seen over Paris on the 24 October, 1870, he writes: (Flammarion 1888, p.767).

> *"....., lumière blanche qui se dissémine à ses bords comme une rosée d' argent."*

so that we would like to think that a possible translation conveying all the shades of meaning of the original Chinese

星隕如雨 should be akin to
"Des *météores* s'effrangeaient en pluie (de feu)"
on the model of a beautiful sentence by Théophile Gautier.

V CONCLUSION: CHINESE AURORAL RECORDS AND THE WOBBLING OF THE GEOMAGNETIC POLE.

By applying the same kind of twofold analysis propounded here to suspected ancient Chinese auroral records, one should be able to correct and complement all the existing auroral lists which, up to now, tried to discriminate true aurorae on mainly intuitive or partial physical considerations. The very nature of the records, put down for astrological reasons and couched in a literary language in which it has always been considered elegant to model oneself on expressions found in the Classics and/or the first two Official Dynastic Histories, as well as the infinite variety of auroral forms of which even a trained observer can often discover new ones he never saw before, make the two criteria probed here probably the only means of some practical value we have to compile auroral catalogues reliable enough for modern scientific investigation.

This kind of compilation shows then that in the most ancient Chinese auroral records the metaphor predominantly used to describe the Aurora Borealis was that of a Dragon. This image, though found in later records, is far less important than we would have thought, had we assumed it to maintain the same momentum throughout Chinese history. In fact, it soon came to be replaced by the concept of *Qì*[i], "Vapour", which is used in about 80% of all the auroral records compiled in the published catalogues. One might argue that this was a mere consequence of the prevailing Chinese philosophical ideas which used this *Qì* concept as a kind of universal tool in order to analyse all the spectrum of Nature, from matter to energy. But this is contradicted by the fact that the "Dragon aurora" image always reappears in-between long spells of *Qì*-aurorae. This leads us to suspect that the so-called *Qì*-aurorae referred in fact to amorphous aurorae, i.e. aurorae seen at low geomagnetic latitudes, which is corroborated by their predominant colour always recorded as red, or reddish. On the other hand, the "Dragon aurorae" should refer to aurorae of such sharply organized forms that the human eye could recognize in them form concepts analyzable by (or perhaps already pre-existent in) the human mind, hence the dragon records of ancient times. These being so numerous until the end of the Posterior *Hàn* dynasty we must admit that during all of the formative period of the Chinese scientific outlook on Nature, China was in much higher geomagnetic latitudes than now, since structured auroral forms can be seen only in the vicinity of the auroral oval. The spells of *Qì*-aurorae periods marked times when China was in lower geomagnetic latitudes[14]. If we eliminate from this phenomenon the "noise" represented by periods of high solar activity, we can get a proof and even an indirect measure of the wobbling of the geomagnetic pole in historic times.

NOTES

1 One may well suppose that Gaubil, who had studied astronomy under Cassini and Maraldi at the Paris Observatory, was fully acquainted with Maraldi's writings, published in 1716-7, on the Aurora Borealis (Eather 1980, p.53). That was just before he left France to go to China (1721).

2 Both these two indentifications are still matters of discussion today.

3 They must have contented themselves with some kind of intuitive identification process.

4 The Official Dynastic Histories of China complemented, in various proportions, by data taken from Local Chronicles and Korean and Japanese histories.

5 Of which there is ample literature.

6 This is, of course, justified by the known fact that auroral forms have remained the same since the dawn of history.

7 Because of space limitations, I shall skip here the first part of my proof, i.e. the philogical analysis of the text. This I would like to publish later, with full details. I recall here, for the benefit of the reader not familiar with the text of the *Book of Changes* that it consists primarily of sixty-four Hexagrams, each of which is made up of a combination of six Lines, either broken or unbroken. To each of these Lines, numbered 1 to 6 from bottom to top, is appended a very short text which often (but not always) can be analysed into a protasis giving the setting, and an apodosis traditionally interpreted as a portent and which we may omit here.

8 This last sentence belongs, properly speaking, to the ominous part of the Fourth Line. However, it is striking to see that, if we take into account very ancient meanings of the Chinese characters it is made up of, the rendering thus obtained still makes sense in an auroral context.

9 In 1.b. and 2.a. Tromholt quotes in fact an auroral description by Weyprecht (1838-1881); it is not quite clear whether the rest of the description is due to Weyprecht or Tromholt.

10 i.e. the period of duration of the auroral display which had begun a little before 23.00 H.

11 This was precisely the case of the three records made use of by *Keimatsu*

12 This precluded one and the same event to be recorded twice within the same Monography, but, of course, did not preclude it to be recorded a number of times in as many different Monographies of one or several Official Dynastic Histories.

13 It is quite illuminating on this point. In order to stress it, I shall hereafter underline the word "meteor" each time I use it in its old sense.

14 For all of the above I am greatly indebted to Professor *N.Fukushima* who, in the course of many discussions, helped me to clarify my views on the historical and scientific analysis of Chinese auroral records.

a. 慶松 光雄 b. 福島 直

c. 易經 d. 坤

e. 乾 f. 星隕如雨

g. 天文志 h. 五行志

i. 氣

REFERENCES

Biot,E. (1844). "Sur la Direction de l'Aiguille Aimantée en Chine et sur les Aurores Boréales observées dans ce même Pays". CRAS, **19**, 822-829.

Eather,R.H.,(1980). Majestic Lights, The Aurora in Science, History, and the Arts, American Geophysical Union, Washington.

Flammarion,C., (1888). L'Atmosphère, Météorologie Populaire, Paris.
Fritz,H.,(1881). Das Polarlicht, Brockhaus, Leipzig.
Gaubil,A., (1970). Correspondance de Pékin, 1722-1759, Droz, Genève.
Gaubil,A.,(1819). Histoire de l'Astronomie chinoise, depuis le commencement de la monarchie chinoise jusqu'ā l'an 206 avant J.-C., Lettres édifiantes et curieuses, **XIV**, 302-447, Lyon.
Keimatsu,M. (1970+). "A Chronology of Aurorae and Sunspots observed in China Korea and Japan. "Annals of Science, Kanazawa University, **7** (1970), 1-10; **8** (1971), 1-16; **9** (1972), 1-36; **10** (1973), 1-32; **11** (1974), 1-36; **12** (1975), 1-40; **13** (1976), 1-32.
Keimatsu,M., Fukushima,N. & Nagata,T.,(1968). "Archaeo-Aurora and Geomagnetic Secular Variation in Historic Time." Journal of Geomagnetism and Geoelectricity, **20**, 45-50.
Link,F. (1962). "Observations et Catalogue des Aurores Boréales Apparues en Occident de - 626 ā 1600." Travaux de l'Institut Géophysique de l'Académie Tchécoslovaque des Sciences, **173**, 297-392.
Link,F. (1964). "Observations et Catalogue des Aurores Boréales Apparues en Occident de 1601 à 1700." Travaux de l'Institut Géophysique de l' Académie Tchécoslovaque des Sciences, **212**, 501-550.
Needham,J.,(1954). Science and Civilisation in China, **3**, Cambridge University Press, London.
Sung,Z.D.,(1973). The Text of Yi King (and its appendixes), Chinese original with (Legge's) English translation, reed.,Taipei.
Teboul,M.(1985). "On the Proper Use of Ancient Chinese Observational Data; The Case of the Aurora Borealis". In Prospect and Retrospect in Studies of Geomagnetic Field Disturbances, Proceedings of a Symposium Dedicated to Prof. *Naoshi Fukushima*, Held January 25-26, 1985, in Tokyo, pp.229-39. Geophysics Research Laboratory, University of Tokyo.
Tromholt,S.,(1885). Under the Rays of the Aurora Borealis: In the Land of the Lapps and Kvaens, **1**, London.

ARTICLES IN CHINESE.

中國歷史上的极光年表(初稿) Peking, 1975.
中朝日历史上的北极光年表,科技史文集 (1980), Shanghai.
中国古代极光年表 Peking, forthcoming

DISCUSSION

Xu Zhentao (Comments) : The data of ancient Chinese astronomy are written in classical Chinese,which is a very difficult language. It is not so easy to grasp the real meaning of these data.Chinese researchers in different teams are currently working very hard to overcome these difficulties. Now we would like very much to unite our efforts with foreign scholars working in the same fields, so that we can achieve better results, and I think this is the best way towards solving the problems we meet.

2.6 WORK OF ART OF THE HAN DYNASTY UNEARTHED IN CHINA AND OBSERVATIONS OF SOLAR PHENOMENA

Xu Zhen-tao
Purple Mountain Observatory, Nanjing,
China.

I THE SUN WORSHIP IN ANCIENT CHINA

In Chinese ancient times the Sun worship was in vogue. Mainly there were two sorts of the worship rites: 1) for rising and setting Sun; 2) for solar eclipses. Many evidences for Sun worship have been found in Chinese ancient books and bone inscriptions (Xu Zhentao et al.1985).

The everyday worship to the Sun certainly caused spontaneous observations for solar phenomena. Since the ancients only had a limited knowledge they could not understand what were these phenomena and they, however, created many wonderful myths to describe them. These myths about Solar phenomena have vividly been expressed in the works of art of the Han Dynasty recently unearthed. So analysing the works of art we may trace the original phenomena of solar activity.

II THE UNEARTHED WORKS OF ART OF THE HAN DYNASTY

A lot of tombs of the Han Dynasty have been excavated in China. For example, a famous one is the tomb at Ma Wang Dui near Changsha in Hunan province. Many works of art have been found from these tombs. They have three main sorts: 1) silk paintings; 2) stone reliefs; 3) mural paintings. Since times of these tombs are very long ago, silk and mural paintings are little in a good state of preservation. But a lot of stone reliefs are very well preserved. In these works of art there are many expressions on Sun's myths. In this paper we regard them as analysing bases.

III THE MYTH ON "Ri Zhong Wu" AND SUNSPOTS

In chinese ancient books "Tian Wen" and "Huai Nan Zi" a myth has been noted. It said that a crow or a three-leg crow was in the Sun. More literary statement was that a famous shooter Houyi shot down nine Sun while there were ten Sun in the sky at that time and the feathers of crow in the Sun were fallen on the ground. Fig.1 is pictographical expression of this myth. The man who is bending a bow is just the ancient hero Houyi and he is aiming at seventh crow, the embodiment of the Sun. Fig.2 and Fig.3 means respectively "the three-leg crow in the Sun" and "the crow in the Sun". The latter is part of the famous silk painting unearthed at Ma Wang Dui. Still many similar works of art on the crow in the Sun were preserved on silk or stone. On the basis of our examinations, these works of art on Ri Zhong Wu have perfectly reproduced concrete figures of solar phenomena which the ancients saw in the sacri-

Fig.3 The crow in the Sun.

(Hanan Museum 1974)

Fig.1 The myth on Houyi shooting Sun.

(Wu Zhende 1984)

Fig.2 The three-leg crow in the Sun.

(Wu Zhende 1984)

ficial rites and the so-called "the crow" or "the three leg crow" really stand for Sunspots(Xu Zhentao et al.1985).

IV THE MYTH ON "Yong Wu Zai Ri" AND SOLAR CORONA AND PROMINENCES

A myth in book "Shan Hai Jing" said that one Sun will rise while another Sun comes here and all of the Sun are carried by crows i.e. so-called "Yang Wu Zai Ri". Fig.4 from a stone relief of the Han Dynasty just is a vivid display of this myth. The big bird called "Yang Wu" stands for one crow and the large circle on the back of the crow means the Sun. On principle mentioned before we can consider this figure as the image of celestial phenomena seen in sacrificial offering to solar eclipses. A convincing evidence of this view point is Fig.5 in which there are a bird "Yang Wu" and a toad in solar disk. As everyone knows, in ancient China a circle with a toad stood for the Moon. So Fig.5 means the coincidence of the Sun and the Moon, i.e. the total solar eclipse. Thus, it is very possible that the head and tail of the "Yang Wu" stand for the equatorial type corona seen in minimum of solar activity and the wings stand for prominences. The total eclipse of May 28th, 1900 (Fig.6) had a bird shape corona similar with Fig.5. It shows that the ancients considered solar corona as the bird "Yang Wu" and this is rather natural.

The Japanese scientist Saito(1979) investigated a lot of figures of the Sun engraved in ancient structures of the Middle East and demonstrated that the two wings beside the solar disk stand for the equatorial type corona while the tail represents a polar plume. It is very interesting that the ancients of the Middle East also considered the Sun as a bird and this just coincides with the Chinese views.

V THE MYTH ON "Yang Li" AND SOLAR FLARES

The book "Tian Wen" said that the big bird Yang Li died and revived. According to literature examinations Yang Li was another divine bird in the Sun different from the crow in the Sun and someone has proved that the story about Yang Li probably means that the Phoenix died in raging fire and revived(XiaoBin 1979). This myth seems to have spread not far and wide, so work of art on it is hardly discovered. Fortunately, a mural painting of tomb of the Western Han Dynasty has been found in 1957 (Xia Nai 1965). It is one of twelve paintings. From Fig.7 we we can see that besides a dark crow there also is another light bird in the centre of solar disk. It is very possible that the light bird just is Yang Li (Xiao Bin 1979).

From Fig.7 we knew: 1) the light bird located near the dark crow which means a big sunspot group as above; 2) the bird Yaing Li is brighter than the crow and even solar disk. In the light of solar physics, it is very possible and very reasonable that the bird Yang Li represents a strong solar eruption, that is a white-light solar flare.

As far as we know that more 60 white-light flares have been discovered since Carritong's sighting. Thus during the long historical years an

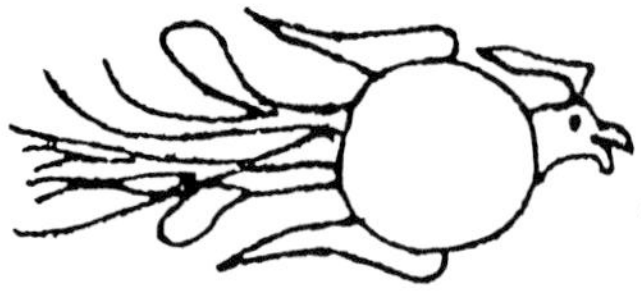

Fig.4 The Sun is carried by crow.
(Wu Zhende 1984)

Fig.5 The coincidence of the Sun and moon.
(Wu Zhende 1984)

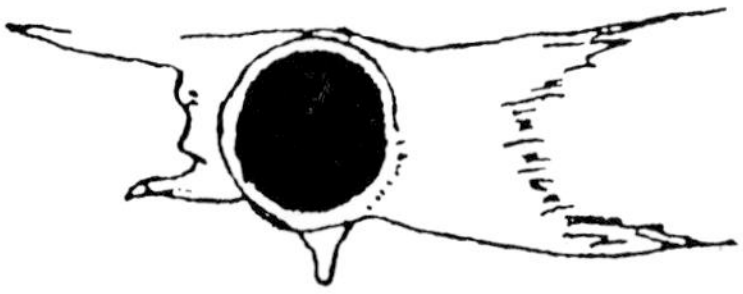

Fig.6 Corona in total solar eclipse of May 28, 1900.

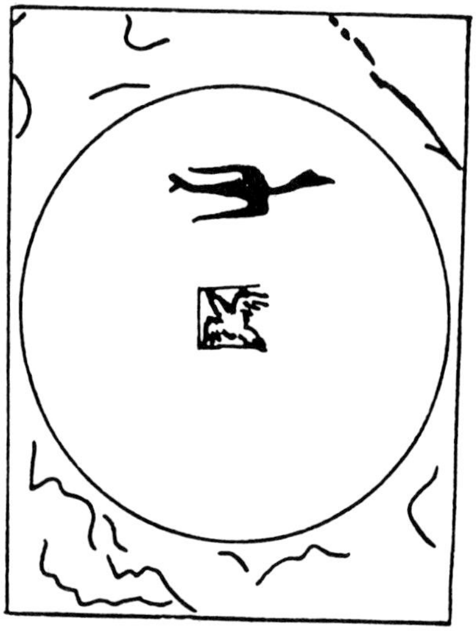

Fig.7 Yang Li and crow in the Sun.
(Xia Nai 1965)

occasional sighting of flare was quite possible. According to excavation report the date of the tomb was about BC48 to BC7 year (Henan 1964). During this period the solar activity had high level corresponding to the Roman Maximum termed by Eddy (Eddy 1976), so easier sightings of solar flares seem to be quite natural. Therefore we can consider Fig.7 as art description of solar flare that the Chinese ancients observed once in a while.

In fact, some descriptions of solar flare observed by naked eye were mentioned in Chinese ancient literatures. For example, "Kai Yuan Zhan Jing"said "there is flame-like gas in the sun" and "one sees red gas as big as a squash jumping in the Sun". Here "flame-like gas" and "red gas" were, in all reason, considered as solar flares witnessed by diligent solar observers (Wang & Siscoe 1980).

VI SUMMARY

The Sun worship led the Chinese ancestors to observe the Sun for a long time. Several phenomena of solar activity were not only offered grand sacrifices but also invented into the wide spread myths which had some reflections in the ancient works of art. Tracing these cultural remains engraved the effects of solar phenomena, we have found that their original embryonic forms right were natural images of the sunspots, solar corona, prominences and even solar flares. Therefore we may affirm that these phenomena of solar activity were initially discovered before the Han Dynasty and their occurrences were so common that the astrologers used them to divine.

REFERENCES

Eddy,J.A., (1976). The Sun since the Bronze Age. In Physics of Solar Planetary Environments, ed. Donald J. Williams, 958. Boulder: American Geophysical Union.

Henan Wenwudui, (1964). Excavation Report of the Western Han Mural Tomb at Luoyang. Kaogu Xuebao, No.2.

Hunan Museum, (1974). Excavation Report on the Han Tombs at Mawangdui. Wen Wu, No.7.

Saito,T., (1979). Solar Corona Appeared in Ancient History Japanese

Wang,P.K. and Siscoe,G.L.,(1980) Ancient Chinese Observations of Physical Phenomena Attending Solar Eclipses, Solar Physics, 66,187. Nature, No.1-2.

Wu Zhende, (1984). The Stone Reliefs of the Han Dynasty. Beijing: Cultural Relic Press.

Xia Nai, (1965). The Mural Constellations of the Western Han Tomb at Luoyang. Kaogu, No.2.

Xiao Bin, (1979). The Silk Painting of Mawangdui and "the Songs of Chu". Kaogu, No.2.

Xu Zhentao et al, (1985). Worship of the Sun and Discovery of Solar Active Phenomena in Ancient China. In Proceedings of Kunming Workshop on Solar Physics and Interplanetary Travelling Phenomena, ed. C. de Jager & Chen Biao, 480-87. Beijing: Science Press.

DISCUSSION

J.A. Eddy : On some occasions sunspots seen with naked eye in ancient China were described as appearing 'like the foot of a crow'. Do you think that this description was an indication of the actual shape of a sunpot group, or, in view of the crow legend, a more allegorical (or poetic) reference ?

Xu Zhentao : I think this is a description of the actual sunspot group. But it is a very complicated problem. I have a paper presented at the International Workshop on Solar Physics (1983), Kunming China, where I reported this problem. I will discuss it with Dr. Eddy in detail.

2.7 RESEARCH ON SCALE AND PRECISION OF THE WATER CLOCK IN ANCIENT CHINA

Quan Hejun
Shanghai Observatory, Academia Sinica
China

The instruments of measuring time have been closely related to astronomy, especially to astrometry. In ancient China, before mechanical clocks were imported from European countries, accurate Water Clocks, main tools for measuring time, had been developed two thousand years ago.

In this paper the sizes (mainly the height) of five single vessel-type Clocks of Western Han Dynasty (206 B.C.-25 A.D.), which were either unearthed in recent years or recorded in some historical documents, are taken as a sample for studying the scaling of indicator-rods and precision of Clock.

I WATER CLOCK OF THE HAN DYNASTY

The Chinese Water Clock had been created long before the Qin and Han Dynasties. At the latest in Shang Dynasty (1711 B.C.-1066 B.C.) the 100-quarter system which could be used to divide a day equally into one hundred ke. In Zhou Dynasty (1066 B.C.-256 B.C.) the Clocks were used in daily life and work, in the imperial court and the army.

At early Western Han, the 100-quarter system was the only system which was used in the Clock. Up to the present, we have found neither any recorded historical documents or picture concerning the Clocks before Han Dynasty, nor any material object. As for the Clocks in Han Dynasty, there are the painting of "Cheng Xiang Fu Clock", the Clock possessed by Prof. Ronggeng, and three Clocks of Western Han Dynasty which were unearthed in recent 25 years from the tomb of Emperor Hanwu, from the tomb of Zhongshan Jing Emperor in the Western Han Dynasty, and from the Inner Mongolia. The type, the size and the structure of these five bronze Clocks are fundamentally similar (Table 1). They are sinking indicator-rods, i.e. outflow Clocks, the four of them are near 20 cm, and the other one is near 10 cm, all of which are smaller than any multivessel Clocks developed later.

TABLE 1
MAIN SIZE OF THE WATER CLOCK IN WESTERN HAN DYNASTY

Name	Depth	Size diameter	Volume	Form
Cheng Xiang Fu	7.5 cun*	5.8 cun		cylindrical
Rong Geng	3.7 cun	1.8 cun		"
Shaanxi	23.5 cm	10.3 cm	2000 cm^3	"
Hebei	20 cm	8.6 cm		"
Inner Mongolia	24.2 cm	18.7 cm		"

*cun, a unit of length (~1/3 decimeter)

II ANALYSIS OF SCALING OF INDICATOR-RODS &PRECISION

The precision of the Water Clock is mainly determined by the stability of outflow. Since the Eastern Han Dynasty (25-220 A.D.) in China, the Clock was developed into the multivessel, which made outflow more even. Since the rate of outflow in the cylindrical single vessel decreases as the water level reduced, so that the indicator-rods which measure time by means of measuring the height of water level should not be scaled with equal intervals.

Indicator-rods were made of materials such as bamboo, wood etc., and they were almost thoroughly rotten when unearthed. Since no historical records could be found, deduction have to be used for analysis. We will discuss on the depth (the height of water capacity) of a single vessel from the viewpoint of unequal scaling. In Fig.1, supose h_0 is the zero level of the water outlet, and

$$h_2 - h_1 = \Delta h_1$$

is the lowest scale in the indicator-rod (1 ke time) (this is a scale with the shortest interval of all the scalings), Δh_1 is reasonably selected in order for it to be able to distinguish by the observer in practical work. The selection of h_1 should not be too low, otherwise the cylinder would be very high (see Table 2)

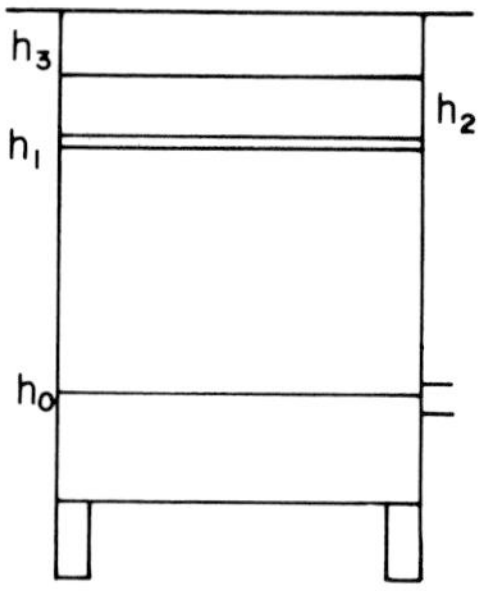

Fig.1

If the shape, the length and the caliber of the water outlet pipe are not considered, by means of fundamental principle of hydrostatics, from h_1, Δh_1 and h_2 the value of Δh_2, i.e., the value of water level reduction of the last but one ke can be calculated, then the water level reduction of the total N ke can be found also

$$\Delta h_2 = h_3 - h_2$$

$$h_3 = \frac{h_2^2}{h_1}$$

since

$$\frac{\Delta h_2}{\Delta h_1} = \frac{\Delta h_3}{\Delta h_2} \ldots\ldots = \frac{\Delta h_N}{\Delta h_{N-1}} = r$$

the formaula of a geometrical series can be used for calculating the water level reduction of the total N ke.

$$H_N = \frac{\Delta h_1(rN-1)}{r-1}$$

In the Western Han period, like the earlier time, the 100-quarter system was used for time measuring. The lengths of the day time and the night were measured respectively from the instants corresponding to sunrise and sunset. The longest time of the day (night) in north of China was approximately above 60 ke. Let us take 100 ke or 70 ke as the whole scale and then calculate the total length of graduation respectively in order to estimate the depth of the Clock. The results are as follows (see Table 2).

Take Δh_1 = 1mm, 2mm
h_1 = 25,40,60,100,150,250 mm

we have the data listed in the Table.

TABLE 2 unit: mm

Δh_1	The lowest level used h_1	Water depth of 70 ke H70	Water Clock depth H70+h_1	Water depth of 100 ke H100	Water Clock depth H100+h_1
1	25	365	390	1240	1265
1	40	180	225 Δ	430	470
1	60	130	190 Δ	250	310
1	100	100	200 Δ	170	270
1	150	85	235 Δ	135	285
1	250	80	330	120	370
2	25	5440	5465	55000	55025
2	40	1180	1220	5220	5260
2	60	525	585	1540	1600
2	100	300	400	625	725
2	150	240	390	565	715
2	250	185	435	305	555

From the above results we can see that if we take Δh_l = 1 mm, the values with Δ in the Table are all roughly similar to the depth of the unearthed Clocks of the West Han period. If the value of h_l is increased by a bit, the depth of the Clock would dramatically increase . The scale of 1 mm is easily distinguished with naked eyes. Because of the continuous moving of the indicator-rod and its nonlinear variation of rate, the scale reading of the single vessel may be set with precision (between half ke and one ke). In the Eastern Han period, the precision of determination of the length of the day or night was merely of 1 ke or so, and so was precision for time measuring. It is not necessary to increase the minimum value of h_l and to enlarge the depth of the Water Clock.

The values deducted from the unequally scalled indicator-rods show that the sizes of Clocks of the Western Han period known at present are much smaller than any of the multivessel. This is why the single vessel discovered are all with small size. (Table 3).

TABLE 3

Name of multivessel-type Water Clock	Size of Water Clock
Yansu (Song Dynasty)	2.15 chi*
Shen Kuo (Song Dynasty)	3.5 chi
Among the people in the Song Dynasty	3 chi
Yanyou (Yuan Dynasty)	75 cm
Qianlong (Qing Dynasty)	97 cm

*chi $\simeq$ 1/3 m

Because of the low precision of the single vessel, the length of the apparent solar day is used as the standard of the length of a day. The length between two successive apparent solar days varies by less than 0.5 sec, and therefore it is impossible for a low precision single vessel to run for a long time with high precision. Thus it is necessary to check it daily or frequently with the apparent Sun, and hence the time scale measured with a Clock is the time of the apparent Sun. Measuring the length of the day-time and night separately is one of the distinguishing features in time measuring in China, since sunrise was taken as a standard of time measuring at ancient ceremony and sacrifice.

In the calendars of ancient dynasties graduations of the day and night in different seasons are given. For instance, graduations for the middle of China are: on the Winter solstice the length of night is 60 ke and on the Summer solstice 40 ke. Accordingly there are more than 180 days between the Winter and Summer solstices, and the difference between the days or nights is equal to 20 ke, that means there is an increasing or decreasing of one ke every nine days. This was the method used in the Qin dynasty. In the early years of the Han dynasty, the Qin's method was still adopted. As variation of the length of the day-time and night at any place is related with declination, an increasing or decreasing of one ke every nine days is approximate. Until Eastern Han Dynasty, in the year 102 B.C., Huo Rong pointed out that this method would not be strict, with which there would be so great a difference as 2-1/2 ke in a year. He suggested a method with which the variation of the length of the day-time and night is changed according to declination, i.e., one ke every 2.4^0 of declination. The value of 2.5 ke must be determined by time measurement with an error, so that the precision of the Water Clock in the Western Han Dynasty would not be higher than one ke everyday. In Eastern Han dynasty, more than two vessel types appeared. To compare the lengths of the day-time and night in different seasons recorded in Si Fen Calendars in Lu Li Zhi (律历志) of the Book of Hou-Han-History with corresponding values it is known that the error given at that time is about half ke. The precision of Clock reached a standard which was worse one ke. In China, the 100-quarter system appeared in a comparatively early period. Comparing the Clock in the Western Han Dynasty with that used in earlier periods, we find that there was no obvious improvement in structure, and therefore the accuracy had not been obviously raised than those before the Qin Dynasty. In China, the Decimal calculating system had an early development history. This was the foundation on which the 100-quarter system was based. The 100-quarter system fulfilled the social activities for a quite long period, but since the Han and Tang feudal societies developed rapidly, more precise time measuring instrument was required in many respects, and that promoted its continuous development. The double-vessel Water Clock was first discovered in the Eastern Han dynasty and after Jin and Tang, the multi-vessel type appeared. This is a different stage of the development of Clock. Until Song dynasty, the multi-vessel typed Clock with steady water level and dividing flows was developed. This is the climax of Water Clock development in China.

DISCUSSION

A.K. Bag : From what time do the most ancient literary evidence on the Water-Clock in China date?

Quan Hejun : They date from the Zhou dynasty.

K.V.Sarma : What were the shapes of the Chinese Water-Clocks?

Quan Hejun : They were cubic, single and composite. In the latter, water flowed from one to the others which were placed lower.

Olaf Pedersen and Mrs. Pedersen
at the Oriental Astronomy exhibition

2.8 GREEK ASTRONOMERS AND THEIR NEIGHBOURS

Olaf Pedersen
Institute for History of Science
University of Aarhus, Aarhus, Denmark

Introduction

In Europe it has been customary to regard the ancient Greeks as our intellectual ancestors. Greek science was seen as the fountainhead from which modern European science ultimately derived both its existence and its characteristic features. This was not a completely empty idea. Each time a modern astronomer mentions a *planet*, the *perigee* and *apogee* of its *orbit*, its periods and their various *anomalies*, he is using so many Greek words. Moreover, until about a hundred years ago the extant works of the Greeks were the earliest scientific texts known to European scholars so that Greek science acquired a unique position in the European mind, and that ancient Greek culture in general became 'classical' and thus an ideal model or pattern for civilization as such. In consequence, the traditional European History of Science became an account of how science arose among the Greeks, how it penetrated into other cultural areas, and how it was sometimes eclipsed and again reborn in one of the so-called 'renaissances' of which European historians are so fond to speak.

That this account was not the whole truth was realised already by the ancient Greeks themselves. They were well aware of being newcomers to the world of science and not afraid of recognising the merits of their predecessors. On the other hand, the actual Greek knowledge of pre-Greek science was always vague and often fanciful although some of the more critical minds of the classical world refused to believe in the more extravagant tales, e.g. the rumour that the Chaldaeans had astronomical records inscribed on bricks and going back to 470 000 years before the Flood. Therefore, classical scholarship paid little or no attention to the few and confused references to earlier science which are scattered throughout the works of the ancient authors.

This situation prevailed until the classical text began to be supplemented by new literary material stemming from the Middle East. The recovery and interpretation of the clay tablets of the 'Babylonians' proved the existence of an astronomical tradition spanning from about 1800 B.C. to the year A.D. 75. These texts revealed a highly developed mathematical astronomy which reached its peak in the last three or four centuries B.C. at the same time as the Greeks began to attack astronomical problems in a serious way. At first sight this new material seemed to underline the unique and original character of Greek astronomy. But slowly some evidence began to appear of a certain interaction between the two cultures and eventually European historians of science realised the necessity of re-examining the history of Greek astronomy in the light of what the Babylonians had achieved.

The Scientific Revolution

It all began in an extremely modest way. The great Homeric epics from about 800 B.C. reveal a very scanty knowledge of astronomical phenomena. A few fixed stars and a couple of constellations are named, and Phosphoros (the Morning Star) and Hesperos (the Evening Star) are mentioned but not identified as the same planet Aphrodite (Venus). The fact that the sun is closer to the horizon in winter times than in the summer is accounted for by saying that the sun-god, Helios, has gone away to the land of the Ethiopians. All phenomena of nature are seen as the results of the voluntary decisions of a great number of divine beings each of whom is responsible for a particular group of phenomena. In consequence, the behaviour of nature could be described in ordinary human language as it had evolved as a means of communication between human beings acting from motives of the same kind as those ascribed to the gods of nature. Thus the account of nature became a discourse about its gods. This mythological approach to nature is found everywhere among ancient peoples. On the one hand, it made nature familiar to man since it was governed on the same principles as human society. On the other, it made the relation between man and nature very uncertain since the acts of the gods were wilful and arbitrary: Who could be certain that the sun-god would return next spring? Here we are at the origin of both magic and divination.

Among the Babylonians, who identified the superior gods with the sun, moon, and planets, astrology occupied a predominant position. From old Babylonian times onwards thousands of astrological omens reveal the existence of professional astrologers who kept watch over the skies and reported notable events to the King in order to enable him to govern, at the same time as they were responsible for the upkeep of the calendar. This connection between astronomical observation and astrological divination was destined to play a decisive role. Over long periods of time it was impossible not to notice a number of regularities in the phenomena of the heavens. So there was only a short step to the actual prediction of future celestial events by more and more sophisticated mathematical methods which in the last four or five centuries B.C. led to procedures for determining new and full moons and synodic phenomena of the planets. This was achieved in a purely phenomenological way and, as far as we can see, without any reference to the cosmological ideas of the structure of the world which must have existed also in Mesopotamia.

Also in the Greek world one must have felt the essential unpredictability of natural phenomena in a mythological discourse as something annoying. This may well have been one reason why the Ionian thinkers in the Greek settlements in Asia Minor began to experiment with a new form of account. This happened in the sixth century B.C. which everywhere in the world seems to have been exceptionally fertile: In India the great Gautama became the Buddha in search of enlightenment, and in Israel the great prophets expelled the many gods of nature in favour of a single God as the creator and upholder of everything according to His unchangeable precepts. What Thales and his followers did was to try to account for all events in nature on the assumption that there are inexorable laws inherent in nature itself of such a force that even the gods cannot resist them. Two centuries after the first tentative efforts of the Ionians the new language of science can be studied in the works of Aristotle where terms like 'cause' and 'effect', 'matter' and 'form', 'substance' and 'accidence', 'chance' and 'necessity' and many others were defined as vehicles for a new type of natural discourse in which uncertain wilfulness was replaced by constant laws which are accessible to the human mind through experience and reason and can be disclosed through natural research. It is obvious that here we are at the roots of that attitude which has ever since been the hallmark of science.

The Initial Period

In a somewhat simplified way it is possible to discern two different tracks along which the early Greeks tried to find the laws which govern the universe at large. The Ionians embarked upon a vast scheme of cosmological investigations embracing everything which exists. This concept was denoted by the word *kosmos* which means 'order', with strong overtones of 'beauty'. Here they made rapid progress. Thales still saw the earth as a flat, round disc surrounded by the ocean. His successor Anaximander gave it the shape of a cylindrical drum, suspended in the middle of a vast space where it remains at rest "because it has the same distance from everything", as he said with a charming application of the principle of symmetry - later generalised into the principle of sufficient reason - which has remained through the ages as a useful heuristic tool for scientific reasoning. Outside the earth Anaximander described the 'stars' or planets as inclined wheels with hollow rims filled with fire which can be seen through a small opening. Other philosophers denied the existence of the sun as a material body, and it was not until after the fall of a large meteorite, in 467 B.C., that Anaxagoras declared the sun to be a red-hot stone larger than Peleponnes. It was also he who realised that the moon gets its light from the sun, an idea which provided him with the correct explanation of eclipses. Physical cosmology has begun to strike roots.

While the Ionians thus tried to account for the universe in terms of interacting material bodies governed by the law of cause and effect, the so-called Pythagoreans - to use Aristotle's sceptical expression - adopted a very different approach. They flourished in the century preceding 400 B.C. when their organisation began to dissolve and other philosophers came to be influenced by their teaching. The Pythagoreans tried to connect the phenomena of nature by means of mathematical relations, in contrast to the Ionian idea of physical forces acting on material bodies. There is no serious reason to doubt the tradition that they discovered the fact that the harmonic properties of the musical scales can be mathematically expressed by ratios between the small integers 1, 2, 3, and 4, and that this led them to establishing the number theory contained in Book VII of the Elements of Euclid.

Since number theory proved to be successful in acoustics it was tempting to try its power also in other fields. According to Aristotle this led to numerological speculations. Since the sum 1+2+3+4 equals the sacred number 10 the early Pythagoreans seem to have assumed the existence of ten celestial bodies, including the earth and the famous 'counter earth', circulating around a central fire. This idea was abandoned by the later Pythagoreans who placed the earth in the middle of the universe with the other bodies circling around it at such speeds and distances that their motions produce the inaudible 'harmony of the spheres'. Likewise they were responsible for the principle that all celestial motions must be circular and uniform with respect to the centre.

Originality and Interaction

In this account there is nothing which could not have been said one hundred years ago. Today we have to make an attempt to distinguish between those ideas which originated among the Greeks and those which they imported from the outside world. In fact, there are many indications that both cosmological and astronomical ideas percolated into Greece, and that they all came from the East. Thus, Thales's idea that the *physis*, or original matter, of everything was water seems to reflect the primordial role ascribed to water in many near-Oriental myths of the origin of the world. Anaximander ordered the heavenly bodies such that the sun is the highest, then comes the moon, next the fixed stars, and finally the five planets closest to the earth; this order is identical with the standard Persian system known from the Avesta. As for the problems of time, Anaximander believed the *physis* to be eternal and performing an eternal motion from which innumerable worlds would emerge and again disappear. Anaximenes taught that there will always be a world, but not the same world, since it becomes different at different times according to certain periods of time. Also Heraclitus is credited with a 'Great Year' of 18 000 years. It is difficult not to connect the belief in definite periods of the world with Eastern ideas, such as the 12 000 years of the Persians, or the even longer periods of both Babylonian and Indian systems. With the Pythagoreans the foreign influence on Greek thought is even more obvious. Thus they thought of the order of the world in dualistic terms such as odd and even, one and many, good and evil, and light and darkness, which are clearly reminiscent of Zoroastrian principles. Un-Greek is also their belief in the transmigration of souls. Finally they also assumed a strictly cyclical theory of time.

All this is sufficient to show that some of the fundamental ideas of the early Greek cosmologists and philosophers had their counterparts in the East and that it is reasonable to assume a direct or indirect transmission. An even stronger suspicion appears if we cross from the realm of ideas to the realm of numerical relations. It is well known that the so-called Metonic period of 19 solar years = 235 synodic months was known in Mesopotamia before Meton's time. Since it is extremely unlikely that Babylonian astronomers would have anything at all to learn from Greek scholars of the fifth century B.C., the inference must be that the Metonic period was imported from Babylon.

Geometrical astronomy

In the beginning of the fourth century B.C. Greek intellectual life was dominated by the philosophy of Plato who was deeply influenced by Pythagorean doctrines and hostile to the naturalistic attitude of the Ionians. But his mathematical proclivities appear in many places, and it is worth noticing that the first known attempt to create a proper planetary theory was made by the mathematician Eudoxus who was, for a time, Plato's collaborator in the Academy in Athens. Eudoxus assumed each of the planets to be carried by a system of spheres, all of them concentric with the earth. In each planetary mechanism the individual spheres had periods and axes of rotation arranged in such a way that the apparent motion of the planet would conform more or less to the observed phenomena, including direct and retrograde motions. However, we have no evidence that the system of Eudoxus was ever used for actual predictions of planetary positions. It also had certain defects. Thus Calippus drew from it the correct conclusion that the sun would move with constant speed on the ecliptic which implied that the four seasons would be of equal length. Towards the end of the fourth century the theory was finally destroyed when Autolycus pointed out that no system of concentric spheres would explain the varying distances of the planets from the earth.

The mathematical theory of concentric spheres was also the basis of Aristotelian cosmology. Aristotle maintained the principle of experience. This meant, firstly, that the mathematical properties of nature were not obtainable from numerological speculation, but only from observations. But secondly, Aristotle wished to treat the celestial world as a material structure like the rest of the universe. In consequence, he replaced the abstract mathematical spheres by spherical shells (also called 'spheres') made of a particular form of celestial matter called aether, an idea founded on Aristotle's belief in the eternity of the world; an everlasting, circular motion of the heavens could not be produced by the four sublunary elements, but presupposed a different kind of matter. Another feature was the finite character of the world. Outside the outermost sphere of the fixed stars there was nothing - not even an empty space. This followed partly from the fact that the notion of an extended vacuum was incompatible with Aristotle's theory of dynamics, partly from the lack of a proper word for abstract 'space' in the Greek language. Thus Aristotle connected celestial mechanics with the laws of physics at the cost of assuming different laws for the celestial and the terrestrial parts of the universe. His cosmology satisfied 'philosophers' and 'physicists' for two thousand years as a crude and easily grasped 'physical' model of the universe, but was useless for astronomers since it suffered from the defects of the homocentric systems in general.

In a more general sense the system of Eudoxus was of immense importance as the first exercise in a type of research in which the Greeks were to become masters - the 'saving of the phenomena' by means of kinematical models of a purely geometrical nature. This led to the development of spherical geometry as a tool for astronomers.

Geometry and astrology

The emergence of geometrical models gave Greek astronomical theory a form which was very different from that of the Babylonian astronomers whose work can be summarily described as an attempt to determine the times and positions of a discrete succession of synodic phenomena without paying much attention to what happened in between. On this background it is understandable that they preferred mathematical tools operating with discrete values of a set of astronomical variables, generated by linear zig-zag or step functions. In contrast to such procedures the geometrical models of the Greeks would work in a continuous way. This fact enabled them to extend the scope and purpose of theoretical astronomy so that it would be possible to solve the problem of determining the position of a celestial body at any given time. It is, therefore, a historical problem if there were intellectual features in the Greek world which would welcome the continuity of geometrical models. Here we may look to astrology which was rooted in the astral religion of Mesopotamia. In Greece the religious background was different and before 400 B.C. there are very few testimonies of astrological beliefs on Hellenic soil, although the way was paved for astrology by the Pythagorean doctrine of world cycles and Plato's description of man as a microcosm closely related to the macrocosm. But gradually Babylonian astrology began to make an impact in the Greek world, and from the third century B.C. astrology spread all over the Mediterranean world. However, this would have been irrelevant to the status of geometrical models if the nature of astrology had not suffered a significant change. Originally astrology was a public affair concerned with omens of future events affecting society as a whole. But from around 400 B.C. astrologers began to predict also the fortunes of individuals from the position of the planets relative to the zodiac at the time of their births, as attested in a cuneiform text from 410 B.C.

This change had important consequences for the astronomical theories on which astrological predictions were made. For whereas the old public or 'judicial' astrology was concerned with the prediction of (mainly) synodic events at certain times in the future, personal 'horoscopic' astrology was faced with the problem of determining planetary positions at arbitrary points of time in the past. It goes without saying that for this purpose a continuously working geometrical model would have a great advantage over the essentially discrete procedures of Babylonian astronomy. There is the difficulty with this assumption that the first Greek horoscopes belong to the second century B.C. and that many Greek astrologers worked by methods derived from Babylonia. However, this does not preclude the idea that a reliable geometrical theory would have been welcomed by astrologers. It only shows that it took a long time before such a theory was forthcoming.

This astrological perspective on theoretical astronomy is, of course, not a sufficient explanation of the Greek preference for geometrical models which also had the advantage of lending themselves to cosmological interpretations; and since from the very beginning of scientific thought the Greeks were seriously concerned with the physical structure of the universe, a geometrical model would to some extent satisfy their growing recognition of the unity of science as an all-embracing account of nature.

Hellenistic Astronomy

The fourth century B.C. had made a new start by investigating a detailed model of planetary motions. This attempt had failed and in the following period Greek astronomers tried to find ways to overcome the defects of the concentric models. Now the scene moved from Hellas and Italy to the new city of Alexandria in Egypt where a kind of Royal Institute with impressive libraries and other facilities for study was able to attract scholars from many parts of the ancient world, among them many astronomers. Here the Aristotelian scheme was generally adopted with that order of the planets which became canonical until the time of Copernicus, defined according to their periods of revolution. Also the absolute finitude of the universe became part of this scheme, despite the attempt made by Stoic philosophers to imagine some sort of emptiness beyond the ultimate sphere. The only serious alternative to Aristotle's cosmology was due to Aristarchus of Samos who, about 280 B.C. proposed a heliocentric model with the earth revolving like a planet in an orbit centered on a motionless sun in the middle of the world. This idea was met with general disbelief and ignored by both 'physicists' and astronomers because the earth had to be without motion for physical reasons, and because the system was not sufficiently provided with numerical parameters to make it possible to calculate positions.

On the other hand, Hellenistic astronomers made a new departure in cosmology by their attempts to determine the sizes and mutual distances of the earth, moon and sun by means of observations and actual measurements. The beginning seems to have been made by Aristarchus who estimated that the sun is 18 to 20 times as far away from the earth as the moon. Somewhat later in the third century B.C. Archimedes found a good value of the apparent diameter of the sun, obtained by direct observation with an instrument constructed for this purpose; and about the same time the circumference of the earth was determined by the Alexandrian librarian Eratosthenes. Such results point to a new awareness of the importance of quantitative observations, often made with new types of instruments such as globes, rings, quadrants, armillary spheres, and others which were developed over all the 700 years of Hellenistic science. Observational astronomy began shortly after 300 B.C. in Alexandria with a certain Timocharis and his successor Aristyllus. Their results seem to have been expressed in degrees and fractions of a degree; if this is not a later reformulation we must conclude that the first Alexandrian astronomers were, in fact, conversant with the Babylonian division of the circle (but not with the sexagesimal system as such) although both Aristarchus and Eratosthenes used earlier methods.

Thus the third century B.C. was a period of transition during which at least some features of Babylonian science were transmitted to the Greek world. This is confirmed by the so-called 'Eudoxus Papyrus' from the beginning of the second century B.C. Its unknown author tried to determine the variation of the length of daylight over the year. He used an example in which the longest day is 14 hours and the shortest 10 hours which were the standard values for Alexandria. In his solution he made use of a linear zig-zag function and it is difficult not to see here an unmistakable Babylonian influence, although the author used old-fashioned Egyptian unit fractions of an hour instead of minutes. Around

150 B.C. the Alexandrian astronomer Hypsicles dealt with a similar problem by another application of one of the mathematical tools of the Babylonians.

Thus Hellenistic astronomy showed both originality and dependence long time before the theoretical problems of planetary motion were attacked in a serious way. However, around 200 B.C. the brilliant mathematician Apollonius of Pergae brought planetary theory out of the deadlock in which Eudoxus had left it. He did not question the old dogma that celestial motions must be circular and uniform with respect to the centre, but investigated two geometrical models in which such motions would, nevertheless, show apparent anomalies to the observer. In the excentric model the observer is outside the centre of a circle on which the planet performs a uniform motion. In the epicyclic model the observer is at the centre of a circle (later called the deferent) on which the centre of an epicycle moves uniformly, while the planet moves uniformly on the epicycle. Apollonius was also able to state that the two anomalies must be identical under certain conditions, and was even able to show how one of the models could be transformed into the other by a geometrical inversion. This revealed that one and the same phenomenon might be 'saved' by different geometrical models; thus it was impossible to draw cosmological conclusions regarding the structure of the world from a geometrical model, a fact which gave rise to fervid discussions among supporters of a 'physical' or a 'mathematical' ideal of theoretical astronomy. On the other hand, there is no doubt that Apollonius's investigations stimulated a new interest in geometrical models of various types in which the resulting anomaly would depend on the choice of the parameters and the sense in which the circular motions were performed. These remarkable achievements owed nothing to the Babylonians, but represented a typical Greek preoccupation with theories shaped in a geometrical form.

The Acme of Greek Astronomy

Although the astrologer Vettius Valens (about A.D. 160) maintained that he had used lunar tables by Sudines, Kidenas (two Babylonian astronomers) and Apollonius, there is no real evidence that the latter put his geometrical models to practical use with parameters drawn from empirical data. This important step was taken by the greatest of all ancient astronomers, Hipparchus (about 160 to 125 B.C.). Almost all of his many works are lost and the results of his achievements can only be inferred from passages in Ptolemy. Extant is his Commentary on a Phainomena by Aratus from about 280 B.C. in which he surveyed the constellations with a wealth of information about positions of several hundred fixed stars, given in various ways which show that Hipparchus had not systematised his celestial frames of reference; in particular it is worth noticing that he only gives ecliptic longitudes for two stars, and no latitudes at all. That he placed the zero point of the ecliptic at Aries 0° shows that here he was independent of the Babylonians who put it at Aries 8°. Of interest is also his solution of the problem of how in each geographical 'climate' the length of the longest day is connected with the solstitial altitude of the culminating sun; here he applied a purely algebraic method. In other problems he used geometrical methods, and everything points to the conclusion that he knew how to apply stereographic projection to solve spherical problems, thereby laying the foundation for the later theory of the astrolabe.

Even more remarkable was Hipparchus's introduction of trigonometrical methods in astronomical calculations by means of a (now lost) Table of Chords which enabled him in principle to solve all problems of plane triangles by calculation and was the immediate basis for his epochmaking work on the theory of the sun. The unequal length of the four seasons implied an annual anomaly of the apparent motion of the sun which made it impossible to use a simple theory of uniform motion on a concentric orbit, so it was natural to try the eccentric (or the equivalent epicyclic) model of Apollonius. The problem was whether this model could be provided with proper parameters, and whether it would then simulate the apparent motion of the sun with sufficient precision. The fundamental parameter was the length of the solar year which Hipparchus here took to be the Calippic year of 365 1/4 days. The length of the spring and summer seasons then led to the length of the corresponding arcs described by the sun on the eccentric circle. From these data and by means of a trigonometrical calculation Hipparchus was able to determine the apsidal line of the orbit and the ratio of the eccentricity and the radius of the orbit. There can be no doubt that he used the model to calculate a table of solar positions throughout the year, and that he found it satisfactory.

This transformation of the abstract model of Apollonius into a practical theory was a major step forward in the history of theoretical astronomy. Among other things it seems to have taught Hipparchus that an eccentric, circular orbit can be determined from only three data. This enabled him to attack the problem of the motion of the moon from its position at three different eclipses, these positions being obtainable from the times of the eclipses by means of the solar theory. The lunar theory of Hipparchus used a model with an epicycle on a concentric deferent. In order to establish its parameters it was necessary to know the different mean periods or the equivalent mean daily motions of the moon in longitude, anomaly, elongation and latitude. According to Ptolemy, Hipparchus obtained these data by calculation from "observations made by the Chaldaeans and in his [own] time", and it is certain that Hipparchus utilized several Babylonian period relations, such as 251 synodic months = 263 anomalistic months and others, for two purposes, viz. to establish a reliable eclipse period in order to be able to check Babylonian eclipse records, and to find precise values of the lunar periods mentioned above. This led him to a value of the length of the synodic month which differed from the Babylonian value, 1 synodic month = 29;31,50,08,20 days, only by a fraction of a second. Thus the extremely precise parameters of the Babylonian lunar theory became an essential element of Hipparchus's work. There is no doubt that Hipparchus finally obtained a lunar theory which worked very well at new and full moons and enabled him to predict eclipses.

Out of the lunar theory came at least one unexpected discovery of the first importance. During two lunar eclipses in 146 and 135 B.C. Hipparchus determined the position of Spica to be roughly 6^{0} west of the autumnal equinox. Now in 294 B.C. Timocharis had observed an occultation of Spica and in 283 B.C. a conjunction between the same star and the moon. The dates of these events gave the longitude of the moon by means of the theory, with the result that relative to the equinoctial points Spica had moved about 2° to the east in the 150 years between the two sets of observations. According to Ptolemy this was the way in which the precession of the fixed stars, or their secular motion in longitude, was discovered. Modern attempts to trace it back to the Babylonians have not met with scholarly agreement.

In the solar theory Hipparchus had assumed a sidereal year. The discovery of precession made it imperative to find a value for the tropical year the precise definition of which is first found in Hipparchus's lost work On the Length of the Year. His result was about 6 minutes too long, a fact of far-reaching consequence for European time-reckoning.

Hipparchus's significance for the history of astronomy cannot easily be overestimated. He gave theoretical astronomy a new start by his demonstration that it was possible to fit geometrical models to observational facts, a lesson which astronomy has never since forgotten. In this he was greatly assisted by his knowledge of Babylonian data; so also in his case we must admit the importance of what the neighbours of the Greeks had achieved, even if his own genius undoubtedly was of the Greek type.

The Final Effort

The 250 years which separated Hipparchus from Ptolemy seems to have been a rather stagnant period. From this interval is a textbook or Isagoge by Geminus who adopted Hipparchus's theory for the sun, but ignored his value of the length of the tropical year and was unaware of the discovery of precession. For the anomalistic month he used a value taken from Babylonian astronomy, and when he calculated the motion of the moon in anomaly he left the geometrical model on one side in favour of an arithmetical scheme; it was obviously derived from the Babylonian system B and provided with Babylonian parameters. All this shows that the influence from the East was increasing and able to compete with the proper Greek tradition. On the other hand some real progress was made in spherical astronomy by the work of Menelaus who defined the concept of a spherical triangle and applied trigonometrical methods to the geometry of the sphere. This opened the whole field of spherical astronomy to numerical calculation of values which could be directly compared with the results of observations, and inaugurated a new era marked by the name of Ptolemy.

Unlike Hipparchus Ptolemy was not a full time astronomer. His purpose was to write comprehensive expositions of all branches of 'applied mathematics', such as astronomy, astrology, geography, cartography, optics and harmonics (the theory of music). That he stands out as the last of the great astronomers of Antiquity is understandable since his Almagest superseded most earlier works. But it must be borne in mind that the Almagest was no report on Ptolemy's own research. It was rather a manual or textbook, written in order to initiate advanced students in the methods of Greek astronomy, and in particular to show how one may construct geometrical models designed to save a great variety of phenomena, and how to provide them with numerical parameters derived from observations.

In the Almagest Ptolemy obviously wished to present solutions to problems which Hipparchus had left unanswered. According to Ptolemy, Hipparchus did not even make a beginning in establishing theories for the five planets, for although he had collected and systematised a number of observations of planetary phenomena he found that these were not in agreement with the hypotheses of the astronomers of that time. But no manual of astronomy could be complete if it ignored the motions of the planets, and there is no doubt that Ptolemy saw it as his task to supplement Hipparchus by trying to construct the missing planetary theories. This was an enormous programme and full of unforeseen difficulties. But first and foremost it was necessary to choose a well defined approach to mathematical astronomy as such, and here Ptolemy consistently discarded the algebraic methods of the Babylonians in favour of the geometrical models of the Greeks. Like his predecessors he made freely use of Babylonian parameters, period relations and records of observations; but the theoretical methods of the East left no trace in his work, neither in his reform of the lunar theory of Hipparchus, nor in his own account of the motions of the five planets which he described by an ingenious combination of epicyclic and eccentric circles the details of which we must here leave out of consideration.

Conclusion

This brief and incomplete survey of some of the main lines of Greek astronomical thought has indicated some of the points at which the traditional picture of the self-contained character of Greek astronomy has undergone revision. We have to admit that European astronomy is not indebted to the Greeks alone. Although they certainly used them in their own characteristic way, they themselves drew freely on observations and parameters of their predecessors in the land between the rivers. In consequence we must abandon the old idea of Graeco-European science as unique and independent. We know already something about the Greek debt to the Babylonians. To investigate the possible, but no doubt existing, relations between the scientific worlds of Greece and India is another task which may well prove to be one of the more fascinating fields of study for both present and future historians of astronomy.

Shigeru Nakayama and Michel Teboul
(Front two from left)

3
Astronomical elements and planetary models

Is it not wonderful in any case, that the most modern and accurate parameters help us to discover and appreciate better than ever before the very accurate Indian observations made nearly 1500 years ago ?

- R.Mercier (p 102)

In any event, there is probably no better place to study comparative science than in Oriental Astronomy. Nevertheless, the field poses formidable problems, foremost being the variety of languages involved. It does no good simply to know Chinese and no other language. We need Chinese scholars who know Sanskrit, Arabic scholars who read Chinese, Indian scholars who command Greek, and so on.

- O.Gingerich (p.274)

3.1 REVISION OF YUGAS AND YUGA-CONSTANTS IN INDIAN ASTRONOMY

K.V. Sarma,
Adayar Library and Research Centre, Madras-20,
India.

Among the historians of Indian astronomy, John Bentley seems to be first to stress, nearly two hundred years ago, that *yuga* and *kalpa*, which form the basic time-divisions used for traditional astronomical computations in India, are not historical but astronomically interpolated. Bentley says that a division of time simply into the four *yugas*, viz. Kṛta, Tretā, Dwāpara and Kali, was introduced in 204 B.C. "It appears," as Bentley surmises, "that at, or about this period (204 B.C.), improvements were made in astronomy; new and more accurate tables of the planetary motions and positions were found, and equations introduced. Beside these improvements, the Hindu history was divided into periods, for chronological purposes. ...The period immediately preceding the inventor was called the first, or *Kali yuga*; the second or next, was called the *Dwāpara Yuga*; the third was called the *Tretā Yuga* ; and the fourth, or furthest back from the author, was called *Kṛta Yuga* and with which the creation began. The end of the first period, called *Kali* was fixed by a conjunction of the Sun, Moon and Jupiter, in the beginning of Cancer, on the 26th June 299 B.C. This was called the *Satya Yuga*, or true conjunction, and is the radical point from which the calculation proceeds" (Bentley 1823, p.61-62).

Bentley then asserts that the *kalpa* division of time was introduced in A.D. 538, and adduces a fantastic reason for the innovation. About the epoch commencing with A.D.538, he says: "This epoch is one of greatest importance, ... as it was now that means were adopted by the Brahmins for completely doing away their ancient history and introducing the periods now in use; by which they threw back creation to the immense distance of 1,97,29,47,101 years before Christian era, with a view, no doubt, to arrogate to themselves that they were the most ancient people on the face of the earth.

"The various means or contrivances that were adopted for this purpose will now be explained:- In the first place, they made choice of a period of 4,32,00,00,000 years, which they called the *Kalpa*. This period they divided and subdivided into lesser periods, which, the better to answer their purpose, they called by the same names of the periods of the two former divisions of the Hindu history were designated, (viz. Kṛta, Tretā, Dwāpara and Kali), in order that they might be conceived to be the same". Bentley continues: "Matters thus far settled, the next step was to ascertain by computation, a point of time from which the calculation of the length of the year and the mean motions of the planets should proceed in order to determine the number of revolutions in each *Kalpa*, preparatory to their application to astronomical purposes. The only point of time they could find to answer this purpose was the 18th February, in the year 1612 of the Julian period, and this point they made the commencement of the *Kali Yuga*, of the 28th *Mahā Yuga*, of the seventh Manwantara" (Bentley 1823, p.69-71).

Bentley proceeds to state: "The point of time thus fixed on was found by computation made backwards, which showed that the planets were approximating to a mean conjunction in the beginning of the sidereal sphere commencing with the Lunar asterism Aświnī, on which account it was made choice as the point to proceed from, for, had the approximation of the planets been in any other part of the heavens, it would not have answered their purpose; because their object was to assume the sun, moon and all the planets to be then in a line of mean conjunction in the beginning of Aświnī, or the sidereal sphere, in order that from that assumption, as if it had been an actual observation, they might determine the length of the year and mean motions of planets, sufficiently near the truth to answer their purpose"(Bentley 1823, p.71-72).

Bentley's prejudices and insinuations apart, and also his way of putting the cart before the horse by proposing that the length of the *kalpas* and *yugas* was decided upon first and then only a particular date in the first year of *kalpa* or *yuga* was sought for answering certain specifications, instead of the other way, his line of argument would be clear from the above.

Roger Billard(1971, p.222), in his recent work, *L'astronomie indienne*, and also elesewhere, ascribes the introduction of the current concept of the four yugas to Āryabhaṭa (born A.D. 476). Billard says: "Not only did Āryabhaṭa construct *yuga* upon such beautiful reductions of observations, but I must add that almost certainly the great astronomer is also responsible for the very introduction of the *yuga* speculation into mathematical astronomy.

Prof. T.S. Kuppanna Sastry observes on this point: "All scholars agree that the mean sun and moon are at the zero point of Aśvinī at the above-mentioned Kali epoch. Excepting school No(3), the others are also agreed that the mean planets too are at that point at the Epoch and that the moon's apogee is 90^{0} and the node 180^{0} from that. What are we to understand from this? Are we to think that at such an ancient date as 17/18 February, 3102 B.C. the Hindu astronomers gave this result as got from their observation? Or, was this point of time fixed by some later astronomers as a convenient epoch for starting their calculations? The former alternative cannot be accepted, because the mean sun, moon and planets were not the same but differed widely from one another, nor were they at zero Aśvinī as calculated by modern astronomy for that epoch. Scholars like Bentley first conceived this idea of verification by calculation. Bentley showed that starting from the epoch and working by each *siddhānta*, the error gradually became less and less, until at the time of the later *siddhāntas*, the error became a minimum, as must be expected. Thus, he proved that the second alternative was the correct one, and that the Kali era starting from this epoch was an extrapolated era founded by astronomical *siddhāntins*" (*Sastry 1974, p.34*).

With regard to original Sanskrit texts, normally one cannot expect them to go into the rationale or the justification or otherwise of the traditionally accepted concepts of *kalpa* and *yuga*. It has however been possible to identify a few texts which categorically state about these concepts being of an interpolated nature. For instance, the Kerala astronomer Putumana Somayāji, mentioning the different yuga divisions adopted by different schools, asserts that these yuga-measures have been conceived only as a means of computation to arrive at correct results. Thus, he states:

> *Kalpādīnām pramāṇam tu bahudhā kalpyate budhaiḥ/*
> *upeyasyaiva niyamo nopāyasyeti yat tataḥ//*
> (*Karaṇa Paddhati*, 5.15)

He then proceeds to recount the different schools;

> *Kalpe yugāni sahasram uśanti kecit*
> *tatraikasaptati yugāni pṛthaṅ manūnām /*
> *ādyantayoś ca vivare ca tathaiva teshām*
> *syuḥ sandhayo yugadaśāṃśacatushkatulyāḥ //16//*
> *manavo'tha caturdaśaiva kalpe*
> *'pṛthu'tulyāni yugāni caiva teshām /*
> *triyugāni gatāni sṛshṭitaḥ prāk*
> *parataḥ syuḥ pralayāt tathāhur anye //17//*
> *yugasya daśamo bhāgo 'bho-ga-pri-ya'hataḥ kramāt /*
> *kṛtādīnām pramāṇam tu syāt pakshayor anayoḥ dvayoḥ //18//*
> *kalpe'smin saptamasyāsya vaivasvatamanor yuge /*
> *ashṭāviṃśe Kaliḥ sarvair vartamāna iha smṛtaḥ //19//*

About Āryabhaṭa's concept of the *yuga*, which is different from the above, our author says:

> *kṛta-tretā-dvāparākhyāḥ kaliś caite yugāṅghrayaḥ /*
> *yugāṅghrayas tu kaple 'smin 'dhigāditya'mitā gatāḥ //*
> (*Karaṇa Paddhati* 1.7)

"The measures of *kalpa* etc. have been conceived by the (ancient) authorities differently, for, it is only the result that counts, not the means (*Karaṇa Paddhati*, 5.15).

"Some (like the *Sūryasiddhānta*, Bhāskara II etc.) take the number of yugas in a *kalpa* to be 1000. Each of the 14 manu periods would have 71 *yugas*; between the beginning and end of each of the 14 *yugas*, there are (in all, fifteen) contact periods, each equal to four-tenths of a *yuga*."(16)

"Still others say that the number of *manu* periods in a kalpa is only 14, each having 71 *yugas*, but that 3 *yugas* have passed by before Creation and 3 *yugas* will occur only after Dissolution".(17)

"According to the above two views, the measure of the Kṛta, (Tretā, Dvāpara and Kali *yugas*) are in the proporation of 4, 3, 2, and 1 tenth parts of the (*catur-yuga*)".(18)

"All agree that today, the current *yuga* is *Kali* in the 28th *(catur-)yuga* of the 7th *manu* (viz Vaivasvata-manu) in the present *kalpa*." (19)

"(According to Āryabhaṭa), (the measures of) each of the four yugas, Kṛta, Tretā, Dwāpra and Kali (are equal,being) one-fourth of the *(catur-)yuga*. In the current kalpa, 1839 quarter-*yugas* are gone! (Karaṇa Paddhati, 1.7)"[1].

Whatever be the *yuga*-concept adopted, astronomers stress that the computed result should accord with observation. Thus, another Kerala astronomer, Parameśvara (1380-1460) states, in his *Sūryasiddhāntavivaraṇa,(Shukla 1957, p.21), under 1.66:*

> *grahāṇām atra siddhānām dṛgbhedo dṛśyate 'dhunā /*
> *dṛgbhedahetuḥ ko'tra syād asmābhir iti cintyate //1//*
> *avyavasthā tu kheṭanām bhukter eva hi yujyate /*
> *śaighryam māndyam tathā kalpyam kramād eva gates tataḥ //2//*
> *kālato gatibhedāc ca siddhānta bahudhā kṛtāḥ /*
> *Brahmādyair ity ataḥ siddhām bheo'tas teṣu yujyate //3//*
> *tattatkāle gativaśād anumānena kalpitāḥ /*
> *bhagaṇās tair nijanije siddhānte siddham ity api //4//*
> *dṛśyamāno 'dhunā teṣām dṛgbhedaḥ sṛṣṭikālataḥ/*
> *ūrdhvakālena sañjāta iti kalpyam budhair ataḥ'//5//*

"There is found difference between the (longitudes of) planets as computed and as observed at present. I am discussing as to what could be the cause of the said difference in the observed positions."(1)

"Variation in the planets can be assigned only with regard to their rates of motion. Hence, increased fastness or slowness, as the case may be, shall have to be presumed in the rates of motion of the planets."(2)

"It is, therefore, only proper that a number of *siddhānta* texts have been produced by Brahmā and others on account of their differences (from one another) in the matter of the time (of their production) and the then rates of motion of the planets."(3)

"Hence it is also proper that the number of *yuga*-revolutions are determined through deduction from the rates of motion at specific times, for the different *siddhāntas*."(4)

"It, again, stands to reason to ascribe the differences that are now observed in the planetary positions to their having developed differences in their rates of motion during the passage of time from creation to the present time."(5)

In the verses following, this author has worked out the actual positions of the planets that would have been occupied by them at the beginning of Kali, on the basis of the current rates of their motion and has given the results as zero corrections to Kali beginning.

Another astronomer who has expressed similar views is Śaṅkara Vāriyar (1500-60), who, in his commentary *Yuktidīpikā* (Sarma 1977, p.73) on the *Tantrasaṅgraha* of Nīlakaṇṭha Somayāji, 1.35, states:

kalyādu na niraṃśatvaṃ bhagaṇāder dyucāriṇāṃ /
gatibhedāt tu dṛksiddhās tatraisham syur dhruvās tataḥ//

"At zero Kali, the revolutions etc. of the planets cannot (be taken to) commence from their zero positions, on account of the (subsequent) change of their velocities. Therefore, zero corrections should be set for them as calculated from their currently observed positions."

In his *Karaṇapaddhati*, Putumana Somayāji not only states the necessity of revising, suitably, the *yuga*-revolutions and other constants, but sets out the methods therefor:

grahaṇa-grahayogādyair ye grahāḥ suparīkshitāḥ /
dṛksamās tatsamāḥ kalpe kalpyā vā bhgaṇādayaḥ //5.1//
parīkshitasya kheṭasya tantranītasya cāntaram /
liptīkṛtyārkabhagṇaiḥ kalpoktaiś ca samāhatam //2//
tantrānirmāṇakālasya parīkshāsamayasya ca /
antarālagatair abdai raśicakrakalāhataiḥ //3//
hṛtvā'ptam tantrānītasya grahasyālpādhikatvataḥ /
svarṇam tat kalpabhagaṇe kuryān naisha vidhī raveh //4//

"(The number of) planetary revolutions (taken to constitute) a *kalpa* (according to the *siddhānta* taken up for consideration) should be revised (periodically) so that eclipses and planetary conjunctions computed using those numbers would accord with observation."(1)

"(Towards effecting such revision) take the difference between the true positions of a planet as observed (in the sky) and as computed using the number of revolutions (enunciated in the *siddhānta*) and reduce the difference to minutes. Multiply this by the Sun's *kalpa* revolutions and divide by the number of years between the time of the composition of the *siddhānta* (or other text) at which time it is to be presumed that the computation accorded with observation). Reduce the quotient (which would be in full cycles) to minutes by multiplying it by 21,600, (being the number or minutes contained in a circle)".(2-3)

"The result (which is the correction in terms of revolutions) is to be added to or subtracted from the number of the (currently accepted) *kalpa* revolutions (enunciated in the *siddhānta* or text) according as the true planet determined by computation is less or more (than the observed true planet). This mode of correction is not to be applied to revise the Sun's revolutions (since the basis itself of the correction is the Sun's revolutions). (4) The above discussion is a pointer also to the fact that astronomical science in India had not remained stagnant and static as is generally supposed and often alleged. It had been evolving, in a particular manner, continually in different spheres and at different levels.

NOTES

1 1839 quarter-*yugas* would amount to 6 *manus*, 27 full *yugas* and 3 quarter-*yugas*, the quarter-*yuga* current today being of Kali yuga in the 28th *manu* period, in consonance with the other schools. There is still another school followed by *Saura-Siddhānta* of the *Pañcasiddhāntikā* of Varāhamihira, followed also by the Midnight system of Āryabhaṭa and the *Khaṇḍakhādyaka* of Brahmagupta.

REFERENCES

Bentley, John -(1797). Remarks on the principal eras and dates of the ancient Hindus. Asiatic Researches, 5, 315-343.
-(1799). On antiquity of the Sūryasiddhānta and the formation of the astronomical therein obtained, Asiatic Researches, **8**, 193-244.
-(1823). A Historical Review of the Hindu Astronomy, Calcutta reprinted London 1825.

Billard, Roger, (1977). Āryabhaṭa and the Indian Astronomy: An outline of an unexpected insight, Indian Journal of History of Science, **12**,, 222.

Karaṇapaddhati - Ed. K. Sambasiva Sastri, Trivandrum Sanskrit Series, **126**, Trivandrum, 1937.

Khaṇḍakhādyaka of Brahamagupta (A.D.628).
1 Ed. with Vāsanābhāṣya of Āmarāja by Babuji Misra, Calcutta Univ. 1925.
2 Translated into English, Univ. of Calcuta, 1934.
3 Ed. with Commentary of Pṛthūdakasvāmin(A.D. 864) by P.C. Sengupta, Calcutta Univ. 1941.
4 Critically edited with commentary of Bhaṭṭotpala and English translation by Bina Chatterjee, 2 Vols, World Press, Calcuta, 1970.

Pañcasiddhāntikā of Varāhamihira(d.A.D.587).
Edited with an original commentary in Sanskrit by Sudhākara Dvivedi and English translation by G. Thibaut, Benares, 1889, republished Motilal Banarsidass, Benares 1930, reprinted 1968.

Sarma, K.V.(1977). Tantrasaṃgraha of Nīlakaṇṭha (A.D. 1444-1545), with Yuktidīpikā and Laghuvivṛti of Śaṅkara, edited with introduction and appendices, Hoshiarpur.

Sastry,T.S.K.(1974). The main characteristics of Hindu astronomy, Indian Journal of History of Science, **9**,, 31-44.

Shukla,K.S.(1957). Sūryasiddhānta with Vivaraṇa of Parameśvara (A.D. 1360-1455), edited with introduction, Lucknow University.

I would like to point out that we must be perfectly clear that the driving motive behind the incessant Chinese observations of the sky phenomena was purely astrological in nature. What was paramount was the ability to know in advance important events bearing upon the immediate future of the ruling imperial family.

- Teboul (p.260)

3.2 SYSTEM OF ASTRONOMICAL CONSTANTS IN HINDU ASTRONOMY

Amalendu Bandyopadhyay
Positional Astronomy Centre, Calcutta 700 053, India.

Ashok Kumar Bhatnagar
Positional Astronomy Centre, Calcutta 700 053, India.

Abstract: Astronomical constants such as the length of the solar year, sidereal and synodic periods of revolutions of the Moon and five brighter planets have been computed using the system of astronomy in ancient and mediaeval India and a comparison made with their modern values. The modern values of the Moon's inequalities have been compared with that of the earlier Hindu astronomical reckonings. Also, the Equation of the Centre of the Sun as determined in the period 500 A.D. to 1150 A.D. has been discussed in relation to corresponding modern values.

INTRODUCTION

S.B. Dikshit (1968) has divided the entire history of Indian Astronomy into three periods viz. (1) The Vedic period, (2) The Vedāṅga period and (3) The Siddhāntic period.

In the Vedic period the study of Astronomy does not appear to have been taken up as an independent science. But it appears that some kind of a luni-solar calendar was in use in this period. The year was solar and consisted of 12 lunar months with the inclusion of 13th intercalary month when necessary. The year had 360 days in it. The natural means of measuring a year used to be one complete cycle of the seasons, the number of which used to be six (sometimes five). Spring used to be the first season. Vedic people were able to realise that Moon's period of revolution was shorter than 30 days, and used a primitive luni-solar calendar. The months were lunar, ending on full moon day. Stars and star groups were called *nakṣatra*.

The *Vedāṅga-Jyautiṣa*, belonging to the *Vedāṅga* period, is the earliest work on Indian Astronomy. This work gives rules for framing a five-yearly calendar on the basis of mean motions of the Sun and the Moon. The months ended with a new moon and the Astronomical elements derived from this work are as follows:

Average length of the year	366 days
Lunar month (a lunation)	29.516 days
Moon's sidereal period	27.313 days

These elements are very crude. The error in lunar months would accumulate to about 1 day in 5 years whereas it will be about 4 days in 5 years in case of solar months. The solar measure is less accurate than the lunar one, perhaps because the stars near the Sun are never visible for direct observations.

The *Siddhāntic* period was heralded by the works of Āryabhaṭa I, who was born in 476 A.D. and wrote two books on Astronomy viz *Āryabhaṭīya* and a *Tantra*. The former reckoned the day from sunrise and the latter from midnight. A large number of works on Astronomy were written during this period. The following table gives the names of the notable astronomers and their works along with the years of their composition.

TABLE 1:
ASTRONOMERS OF THE *Siddhāntic* PERIOD AND THEIR WORKS.

Āryabhaṭa I	*Āryabhaṭīya* and another *Tantra*	499 A.D.
Lāṭadeva	Expounder of *Romaka* and *Pauliśa Siddhāntas*	505 A.D.
Varāhamihira	*Pañca-Siddhāntikā* which includes the *Sūrya-Siddhānta*	c. 550 A.D.
Brahmagupta	1. *Brāhma-Sphuṭa-Siddhānta* 2. *Khaṇḍakhādyaka*	628 A.D. 665 A.D.
Lalla	*Śiṣyadhīvṛddhida*	748 A.D.
Vaṭeśvara	*Vaṭeśvara-Siddhānta*	904 A.D.
Muñjāla	*Laghumānasa* and *Bṛhanmānasa*	932 A.D.
Śrīpati	*Siddhāntaśekhara*	1039 A.D.
Bhāskara II	*Siddhāntaśiromaṇi*	1150 A.D.

SIDEREAL AND SYNODIC PERIODS

Table II gives a comparative study of the sidereal and synodic periods of the Sun, Moon and the Planets as given by the various astronomers of the Siddhāntic period. The astronomers of ancient India were successful in determining the synodical periods of the planets with a greater degree of accuracy than in their determination of the sidereal periods. In the case of the Moon, however, the Indian astronomers were successful in their determination of the value of the lunar month which is now found to be correct within a fraction of a second, although the true position of the Moon shows a great divergence, which is mainly due to their neglecting the additional corrections of Muñjāla and Bhāskara.

MOONS MOTION:LUNAR INEQUALITIES

The modern value of the Moon's inequalities up to the first five terms is stated as

$+377'.3 \sin g' + 12'.8 \sin 2g'$	Equation of centre
$+ 76'.4 \sin(2D-g')$	Evection
$+ 39'.5 \sin 2D$	Variation
$- 11'.2 \sin g$	Annual Equation

where g' and g are the mean anomalies of the Moon and the Sun measured from their respective perigees, and D is given by "mean Moon - mean Sun".

The above expression with the three principal terms may be put into the form

$$-300'.9 \sin g_1 - 152'.8 \cos(D-g_1) \sin D + 39'.5 \sin 2D$$

where g_1 being measured from the apogee is given by $g'+180^{\circ}$. In the earlier Hindu astronomical reckonings, we come across only the first term of the above lunar inequalities. But when we come down to the time of Muñjāla (932 A.D.), we get the term of the second inequality in the form

$$-144' \cos(\odot - \alpha) \sin D,$$

Where α stands for the lunar apogee. This, it will be seen, is exactly the modern form of the evection as combined with a part of the equation of apsis shown above. Srīpati also tried to express the second inequality after the manner of Muñjāla but his constant is equal to 160' instead of Muñjāla's 144', while the correct value is 153'. The third inequality of the Moon known as 'Variation' was used by Bhāskara II alone, his constant being 34' instead of 40' which is taken to be the correct value. The orthodox almanac makers of India, most of whom follow only the *Sūrya-Siddhānta* for their calculations, do not take account of any of these corrections except the first term, viz., $-300'.9 \sin g_1$ in finding the Moon's place. As a result they are giving in their almanacs the positions of the Moon which often differ from the actual position by as much as 3 degrees of arc (or about 6 hours in time). It is, however, interesting to note that the additional terms of both Srīpati and Bhāskara vanish when D equals 0^{0} or 180^{0}, which means that at new Moon and full-Moon the position of the Moon is given rather correctly by the above mentioned first term only, except for the residual discrepancy caused by the 'annual equation' which also appears to have been compensated at syzygies by amalgamating the term with the 'equation of centre' of the Sun as shown below.

EQUATION OF CENTRE OF THE SUN

As regards the equation of centre of the Sun, the modern value of its principal term is $115'.2 \sin g$; in 500 A.D. it was $119'.1 \sin g$, slightly greater than its present value. The corresponding term

TABLE II.

SIDEREAL PERIODS IN DAYS

Planet	ARB	BS	KK & SSV	VS
Sun	365.258681	365.258438	365.258750	365.258694
Moon	27.321672	27.321667	27.321674	27.321670
Mercury	87.96988	87.96992	87.96999	87.96971
Venus	224.69814	224.69794	224.69818	224.69853
Mars	686.99974	686.99793	686.99987	686.99857
Jupiter	4332.27217	4332.24009	4332.32058	4332.31992
Saturn	10766.06465	10765.81524	10766.06670	10765.77125
Moon's apogee	3231.98708	3232.73410	3231.98769	3232.09313
Moon's asc.node	6794.74951	6792.25396	6794.75080	6794.39868

	SSN	SSN with bija corrections	PTM	MOD
Sun	365.258756	no change	365.246667	365.256363
Moon	27.321674	no change	27.321667	27.321661
Mercury	87.96970	87.96978	87.96935	87.96926
Venus	224.69857	224.69895	224.69890	224.70080
Mars	686.99749	no change	686.94462	686.97985
Jupiter	4332.32065	4332.41581	4330.96064	4332.58892
Saturn	10765.77307	10764.89172	10749.94640	10759.22653
Moon's apogee	3232.09367	3232.12016	3231.61655	3232.58853
Moon's asc.node	6794.39983	6794.28281	6796.45587	6793.45994

SYNODIC PERIODS IN DAYS

Planet	ARB	BS	KK & SSV	VS
Moon	29.530582	29.530582	29.530587	29.530583
Mercury	115.87833	115.87843	115.87852	115.87803
Venus	583.89746	583.89675	583.89758	583.90008
Mars	779.92103	779.92225	779.92117	779.92260
Jupiter	398.88950	398.88948	398.88917	398.88911
Saturn	378.08595	378.08599	378.08602	378.08632
Moon's apogee	411.79741	411.78498	411.79749	411.79571
Moon's asc.node	386.00899	386.01677	386.00906	386.01013

	SSN	SSN with bija corrections	PTM	MOD
Moon	29.530588	no change	-	29.530588
Mercury	115.87801	115.87815	115.8786	115.87748
Venus	583.90018	583.90277	584.0	583.92137
Mars	779.92427	no change	779.9428	779.93610
Jupiter	398.88918	398.88837	398.8864	398.88405
Saturn	378.08639	378.08747	378.0930	378.09190
Moon's apogee	411.79578	411.79535	-	411.78470
Moon's asc.node	386.01020	386.01058	-	386.01056

Abbreviations

ARB	*Āryabhaṭīya*	VS	*Vaṭeśvara Siddhānta*
BS	*Brāhmasphuṭa Siddhānta*	SSN	*Sūrya-Siddhānta (New)* as available at present
KK	*Khaṇḍakhādyaka*		
SSV	*Sūrya-Siddhānta* as known to Varāhamihira	PTM	Ptolemy
		MOD	Modern

of the Indian astronomers is, however, 131' sin g, which, as is apparent, does not compare favourably with the correct value. But when we take that the Indian Astronomers were more interested in correctly determining the difference between the longitudes of the Sun and the Moon particularly at new-Moon and full-Moon, it is possible to find out an explanation for the above discrepancy. Taking the values depending on g only in the solar and lunar inequalities, it is observed that

$$\odot - ☾ = 119'.1 \sin g - (-11'.6 \sin g) = 130'.7 \sin g$$

which is exactly the value adopted by the Indian astronomers.

REFERENCES

Bentley, J. (1825). A Historical view of the Hindu Astronomy, Calcutta: The Asiatic Society.

Bose, D.M., Sen, S.N. & Subbarayappa, B.V. (1971). A Concise History of Science in India. New Delhi: Indian National Science Academy.

Burgess, E. (1935). Translation of the Sūrya-Siddhanta. Calcutta: University of Calcutta.

Dikshit, S.B. (1968). Bharatiya Jyotish Sastra, Parts I & II (History of Indian Astronomy). Calcutta: Positional Astronomy Centre.

Kaye, G.R. (1924). Hindu Astronomy. Calcutta: The Archaeological Survey of India.

Saha, M.N. & Lahiri, N.C. (1955). Report of the Calendar Reform Committee. New Delhi: Council of Scientific & Industrial Research.

Sen Gupta, P.C. (1939). Hindu Astronomy. Belur Math, Calcutta: The Cultural Heritage of India, Vol.3.

Shukla, K.S. (1957). The Sūrya-Siddhānta. Lucknow: University of Lucknow

Shukla, K.S. (1960). Bhāskara I and His Works. Lucknow: University of Lucknow.

Shukla, K.S. (1976). Āryabhaṭīya of Āryabhaṭa. New Delhi: Indian National Science Academy.

Thibaut, G. & Dvivedi, S. (1968). The Pañcasiddhāntikā of Varāha Mihira. Varanasi: Chowkhamba Sanskrit Series Office.

DISCUSSION

S.D.SHARMA (Comment) : David Pingree has claimed that Munjal's correction is Arabic. Since the Equations of Centre of Indian Tradition are of different origin and were computed using two eclipses with sun at its apogee and moon at 90^0 from its Apogee and vice versa. The correction by Munjal is a hybrid of 1st equation of centre of moon of Indian tradition; it cannot be of Arabic origin as Arabs used $2^0 23'$ as equation of the centre for sun. This is found to have been used by Almijisti. Note that sun's equation in Indian tradition has annual variation subtracted from it $1^0\ 55' - (-15') = 2^0\ 10'$.

Abridged Armilla. At the beginning of the Yuan Dynasty, astronomer Guo Shou-jing simplified the complicated ancient instrument so as to separate the circles of the horizontal coordinate system from those of the equatorial system. It can avoid the obscuration of the observable sky region by the many rings of the old instrument. It was made in 1437, the second year of the Zheng-Tong Reign of the Ming Dynastry.

3.3 INDIAN CALENDARS

S.K.Chatterjee
A-9/1 Vasant Vihar, New Delhi 110 057,
India.

VEDĀṄGA JYAUTIṢA CALENDAR

The first treatise on calendric astronomy was compiled C1300 B.C.and is known as "The *Vedāṅga Jyautiṣa*. It gives rules for framing calendar covering a five-year period, called a *'Yuga'* . In this *yuga*-period calendar, there were 1830 civil days, 60 solar months, 62 synodic lunar months, and 67 sidereal lunar months. The calendar was luni-solar, and the year started from the first day of the bright fortnight when the Sun returned to the Delphini star group. Corrections were made, as required, to maintain this stipulation to the extent possible. The *Vedāṅga* calendar was framed on the mean motions of the luminaries, the Sun and the Moon, and was based on approximate values of their periods. *Vedāṅga Jyautiṣa* calendar remained in use for a very long time from C 1300 B.C. to C 400 A.D. when *Siddhānta Jyautiṣa* calendar based on true positions of the Sun and the Moon came into use and gradually replaced totally the *Vedāṅga* calendar.

THE SŪRYA SIDDHĀNTA

On the dwindling of *Vedāṅga Jyautiṣa* system, many books on *Siddhānta Jyautiṣa* appeared, but the best treatise on the subject and also one that is followed now is named as *Sūrya Siddhānta*. It is not known who was the author of this famous book, and when it first came into use. It completely replaced the Vedāṅga Jyautiṣa calendar which ruled the calendric system of this sub-continent for more than 1500 years.

The *Sūrya Siddhānta* astronomers devised rules for framing the calendar on the basis of true positions of the Sun and the Moon as against their mean positions used earlier. They calculated the year on the basis of sidereal or *nirayana* system as opposed to tropical or *sāyana* system followed by the Gregorian calendar. The word *'nirayaṇa'* means 'with no motion' and the word *'sāyana'* means 'with motion', and this refers to the vernal equinox which is constantly retrograding, the present rate being 50".3 per year. The fixed point from which all *nirayaṇa* calculations are made is that which is opposite the Star Spica (α-Virginis), called Citrā in the Indian language. In other words, from this point, the longitude of Citrā or spica is 180^0. For assigning a precise position of this point, the Indian Astronomical Ephemeris has adopted its tropical longitude as $23^0 15'00''$ on 21 March 1956. The angular distance in longitude of this fixed point from the vernal equinoctial point on 1 November 1985 was $23^0 39'.3$

The Zodiac belt, which is called *rāśi cakra* of this calendric system, is divided as usual into 12 equal parts starting from the aforementioned fixed initial point, and each part or zodiac sign spanning 30^0 along the ecliptic is known as *rāśi* , and the names of the twelve *rāśis* counted from the beginning are (1) Meṣa (2) Vṛṣa (3) Mithuna (4) Karkaṭa (5) Siṃha (6) Kanyā (7) Tulā (8) Vṛścika (9) Dhanuṣ (10) Makara (11) Kumbha and (12) Mīna, corresponding English names for the above *rāśis* or zodiac signs are: (1) Aries (2) Taurus (3) Gemini (4) Cancer (5) Leo (6) Virgo (7) Libra (8) Scorpio (9) Sagittarius (10) Capricon (11) Aquarius and (12) Pisces. But the difference is that in the English or Western system, these zodiac signs are reckoned from the moving vernal equinoctial point and not from a fixed point in the sky, and hence in the two systems the zodiac signs of same name do not coincide, and the Western zodiac signs are $23^0 39'$ on the west of those of the same name in the Indian system.

CALENDARS FOLLOWED UNDER SŪRYA SIDDHĀNTIC SYSTEM

In India, apart from the Gregorian calendar which is used by the government and commercial houses, mainly three different calendars based on *Sūrya Siddhānta* system, are in use today. One is a solar calendar, and the two others are luni-solar ones. The main difference between the two luni-solar calendars is that in one the lunar months are new-Moon ending, and in the other these are full-Moon endings. These three different calendars are described below.

1 SOLAR CALENDAR

The solar calendar used in India is on *Sūrya Siddhānta* system, and the length of the year is sidereal (*nirayaṇa*) as opposed to tropical (*sayana*), which is used in framing the Gregorian calendar. The length of the sidereal and tropical years in mean solar days are respectively 365.25635 and 365.24218, and hence the former is longer than the latter by 0.01417 day or by $24^m 20^s$. There are the usual twelve months in a year, and these are named serially as follows: (1) Vaiśākha, (2) Jyaistha (3) Āṣāḍha, (4) Śrāvaṇa (5) Bhādra, (6) Āśvina, (7) Kārtika, (8) Agrahāyaṇa, (9) Pauṣa, (10) Māgha, (11) Phālguna, and (12) Caitra. These 12 months are linked respectively with the 12 *rāsis* or zodiacal signs, starting from Meṣa, mentioned earlier, that is, Vaiśākha is linked with Meṣa, Jyaistha with Vṛṣa and so on. The meaning is that the length of months is determined by the time taken by the Sun to traverse the concerned *rāśi* with which the month is linked. For this reason, in some regions months are named as *rāśis*.

The actual time taken by the Sun to traverse the *rāśis* varies from 29.45 to 31.45 days as per Kepler's law applied to the elliptical orbit of the earth where the Sun is located in one of its foci. Again the ingress of the Sun from a *rāśi* to the next (*saṃkrānti*) may take place at any time of day or night but the calendar day of the month, as followed under the traditional system, starts with sunrise. Therefore, a rule or convention is required to be followed for dertermining the day when the month will start. There are, however, four conventions followed in four different regions,namely, Bengal, Orissa, Tamil Nadu and Kerala, and this has given rise to the position that the month may commence on the day of *saṃkrānti* (ingress of the Sun to the next *rāśi*), or on the day following,or sometime on the day after. The effect of all this is that the months of the solar calendar used in *pañcāṅga* (Indian almanac) may vary from 29 to 32 days. Further, in the same year the same months may not have the same length in all regions, and again in the same region the length of the same month may vary from year to year because months have no fixed number of days.

The solar calendar for counting the days of the months and consequently the year is followed in the states of Tripura, Assam, Bengal, Orissa, Tamil Nadu, Kerala and partly in Punjab and Haryana. In all other states, the luni-solar calendar is followed for calendrical purposes. Generally, however, the northern states follow the full-Moon ending pattern while the southern states follow the new-Moon ending one. The details of these two types of luni-solar calendar are described below:

2 NEW-MOON ENDING OR AMĀNTA LUNAR CALENDAR

The months of this calendar are counted on the time period of one new Moon to the next, and are named after the solar months in which the initial new Moon from which the lunar month starts, occurs. In both new-Moon and full- Moon ending lunar calendars, the year starts from Caitra while in the solar calendar it starts from Vaiśākha, the month following Caitra when the Sun enters Mesa *rāśi*, that is, sidereal sign of Aries.

The lunar month being synodic, the Moon moves through 360^0 from one new Moon to the next in relation to the Sun. The period of time the Moon takes to gain successively 12^0 over the Sun is known as *tithi*, and there being thus 30 *tithis*, in a lunar month, 15 in the waxing or bright period (*Sukla pakṣa*) and 15 in the waning or dark period (*Kṛṣṇa pakṣa*) of the Moon. The day number of the month of the lunar calendar is reckoned on the basis of the ordinal number of the *tithi* current on the day at sunrise. Now the motions of the Sun and the Moon not being steady, the duration of a *tithi* varies from 26.78 to 19.98 hours and this results in a *tithi* covering sometimes two sunrises or falling between them. When this happens, there is a break in the seriality in the day-number of the days of the month because then the *tithi*-day number will be repeated or omitted.

The mean duration of a lunar month is about 29.53 days and the lunar year thus falls short of the solar year by about 10.9 days. In the luni-solar calendar this discrepancy is made up by inserting an additional or intercalary lunar month called *adhika* month to keep the lunar calendar adjusted with the solar and consequently the lunar months with the seasons. The procedure followed for this adjustment in the lunisolar Jewish calendar is that 7 intercalary lunar months are added at laid-down intervals in 19 lunar years making the total number of days very nearly equal to 19 solar years. The Indian lunar calendar, however, has not adopted the above procedure of adding intercalary months in a mechanical manner. The early Indian astronomers had devised an ingenious astronomical method for determining the intercalary months which is just as accurate. The method followed is that when two new Moons occur in one solar month which is calculated on the basis of exact time taken by the Sun to traverse the concerned *rāśi* then first lunar month commencing with the first new Moon is treated as an intercalary or *adhika* month, and the second one as a normal or *śuddha* month. Both lunar months, however, have the same name as they commence from the same solar month, but the first month has the prefix *adhika* and the second one *śuddha* added to them.

It sometimes happens, though at long intervals of time, which may be as early as 19 years or as late as 141 years, average interval being about 63 years, no new Moon may occur in a certain short solar month like Agrahāyaṇa, Pauṣa, or Māgha, and in that case there will be no lunar month after the name of that solar month and thus forming a void in the seriality of lunar months. This void month is known as *kṣaya* month. When a *kṣaya* month comes about, there always occur two intercalary or *adhika* lunar months in two solar months, one before and one after the solar month in which no new Moon has occurred. One of the intercalary months, normally the first one, is treated as an intercalary or *adhika* month and the second one is treated as normal or *śuddha* month, filling the void of *kṣaya* month and making the lunar year comprise of normal 12 months.

Amānta luni-solar calendar is followed in the states of Maharashtra, Gujarat, Andhra Pradesh and Karnataka. In all states, however, *amānta* luni-solar calendar is used for fixing the dates of religious festivals which depend on lunar calendar.

3 FULL-MOON ENDING OR PŪRṆIMĀNTA LUNAR CALENDAR

The months of this calendar are reckoned on the basis of time period from one full Moon to the next, but is named after the *amānta* (new-Moon ending) calendar commencing a fortnight later. In other words, *pūrṇimānta* lunar month begins a fortnight earlier and ends in the middle of the *amānta* month of the same name. The other features of this calendar are the same as those mentioned for *amānta* calendar.

ERAS

Apart from different types of calendar followed, use of different eras are also in vogue. States following *pūrṇimānta* lunar calendar and the state of Gujarat following *amānta* lunar calendar, use Vikrama era which commence from 58 B.C. All states following *amānta* lunar calendar except Gujarat, use Śālivahana Śaka era commencing from 78 A.D. Śaka Solar era is used by all states following solar calendar, and in addition Bengali San is used in Tripura, Assam, and Bengal, and Kollan era in Kerala. The Kali era commencing from 3102 B.C., used by the pioneer early astronomer Āryabhaṭa, is used throughout in addition.

EVOLVING A STANDARD INDIAN CALENDAR BASED ON NATIONAL TRADITION

The three calendars described above are the principal ones that are in use at present. There exist a number of other different calendars but their use is restricted either to certain areas or to some groups of persons.

The existence of so many different calendars is a legacy of the past political division of the country when several independent kingdoms flourished. Now, attempt is being made to have a standard calendar for the entire country to be used for all purposes. Such a calendar to become acceptable has to keep in mind the traditional system which has been in vogue for many centuries, and at the same time it has to be scientific.

R.Mercier examining books at the
Oriental Astronomy exhibition

3.4 THE MERIDIANS OF REFERENCE OF INDIAN ASTRONOMICAL CANONS

Raymond Mercier
Southampton University, Southampton, S09 5NH, England.

ABSTRACT The canons of Sanskrit astronomy depend on mean motions which are normally postulated to refer to the central meridian of Ujjain. The present work is a statistical analysis of these mean motions designed to discover the optimum position of the meridian, by comparison with modern mean motions. This follows earlier work done by Billard in determining the optimum year. The results confirm that from the time of Āryabhaṭa all the canons were referred to meridians lying well within India, and in many cases clearly identifiable with Ujjain within the statistical bounds.

1. INTRODUCTION

The general idea underlying the research which is summarised in the present paper is hardly original, and indeed is a direct development of that employed so successfully in Roger Billard's l'Astronomie Indienne (Billard (1971)). The medieval mean longitudes of Sun, Moon and planets are compared directly with the corresponding modern means, and the differences are treated by the method of least squares in order to determine values of the year and of the meridian to which the medieval longitudes are referred.

In Billard's researches the meridian was always assumed to be that passing through Ujjain, the central meridian repeatedly referred to in Sanskrit sources. In this way he was able to fix, within narrow limits as a rule, the year when the Indian mean longitudes best agreed with the true configuration.

In the present work not only the year, but also the meridian are allowed to vary in the search for the best fit. There are also improvements in the modern parameters, taking advantage of quite recent developments, and besides the statistical control follows more closely the formal method of least squares.

Apart from a natural confirmation of Billard's results for the optimum year, we have a series of results for the optimum meridian showing that it always lies in India, indeed generally near to Ujjain (longitude 75;46 East of Greenwich). This reinforces very well Billard's general conclusion that with the Sanskrit canons we are presented with a continuous millenium of observational astronomy in India.

All of the canons cited by Billard have been analysed in this way, as well as a number of others, but here there is room only to present a few particularly important ones.

2. MEAN LONGITUDES AND DEVIATIONS

Let the modern mean longitudes be denoted $L_i(t)$, where the suffix runs from 1 to 9:

1. Sun
2. Moon
3. Lunar apogee
4. Lunar node
5. Mercury
6. Venus
7. Mars
8. Jupiter
9. Saturn

All these are tropical longitudes. The precise numerical expressions are those now to be employed in the national ephemerides, following an IAU recommendation (Francou, e.a., 1983). These expressions are functions of Terrestrial Dynamical Time (TDT) which differs from Universal Time by the quantity ΔT, which allows for the changes in the rate of rotation of the earth; this includes both a steady rate of decrease, and fluctuations. A satisfactory expression for the steady part of ΔT depends necessarily on ancient eclipse records, and this calculation is in no way as accurate as that determining L_i. For many years a formula for ΔT which was determined by Spencer Jones (1939) has been used, although there are theoretical objections (van der Waerden (1961)) to its derivation. Many new formulae have been derived, but they do not agree especially well among themselves, nor do they meet the objections brought by van der Waerden against the older formula. Naturally the uncertainty in the value of ΔT appears directly in the meridian which we determine, but it is unlikely that further revisions of ΔT would lead to alterations in the meridian of more than a few minutes of arc.

In this paper the expressions for $L_i(t)$ include terms as far as t^3 (as taken from references listed by Francou, e.a., (1983)), and in addition a large number of trigonometrical terms expressing the various perturbations (1). These are included for Venus, the Sun, Mars, Jupiter and Saturn, but are effectively important only for the last two, which are affected by the well known resonance.

The medieval mean longitudes $\lambda_i(t)$, where t is simply the Universal Time, are linear expressions. In most cases, the canons which we analyse are already included in Billard's survey, and there one will find the numerical details. As a general rule the Sanskrit canons define sidereal longitudes.

3. THE METHOD OF LEAST SQUARES

We now change the notation slightly, so that t denotes the Universal Time, and TDT as required for L_i will be $t + \Delta t$, with Δt provided by the Spencer Jones formula, *faute de mieux*. The deviation between the modern and the medieval mean longitudes is defined as

$$D_i(t,\phi) = \lambda_i(t-\phi/360) - L_i(t+\Delta t)$$

where ϕ is the longitude of the meridian East of Greenwich.

The method of least squares will be used to determine jointly the optimum estimates t_0 and ϕ_0. For this purpose we must postulate 'true' values of $D_i(t,\phi)$, which may be done in two ways. If λ_i are tropical longitudes, such as we may calculate from the Sanskrit canon when we are also given a model of precession, then the 'true' value of D_i is zero, and then we would seek the minimum of the sum

$$Q = \sum_I D_i^2$$

where I is a selection of values of i. In this case we say that the deviations are 'absolute'. On the other hand, if λ_i is sidereal, the 'true' value of D_i would depend on the rate of precession together with an unknown constant. In this case therefore, which we refer to as 'relative', we can only take the mean of the set D_i as the true value, and calculate the minimum of

$$Q = \sum_I (D_i - (\sum_I D_i)/N)^2$$

where N is the number of deviations included. When the deviations are relative, the number of random variables is then N - 1. Let I be indicated by a sequence of 1's or 0's, indicating whether a particular value of i is included or not: thus (1101 00000) would indicate that the Sun, Moon and lunar node only are included. Further the set I will be augmented by an initial 1 or 0 to indicate whether we are concerned with absolute or relative deviations, respectively. With each determination of the year and the meridian, then, we associate a statistic I, such as (0 1101 0000). This is the same symbol as used by Billard.

The method of least squares has been explained in a suitably general and clear way by van der Waerden (1967). In the following only the briefest summary of the application is possible.

If we write in the neighbourhood of the minimum

$$Q = h_{11}(t-t_0)^2 + 2h_{12}(t-t_0)(\phi-\phi_0) + h_{22}(\phi-\phi_0)^2 + Q_0,$$

then the estimate of the variances of t_0 and ϕ_0 are s_t^2 and s_ϕ^2:

$$s_t^2 = \frac{Q_0}{n-r}\,\frac{h_{22}}{h_{11}h_{22}-h_{12}^2}$$

$$s_\phi^2 = \frac{Q_0}{n-r}\,\frac{h_{11}}{h_{11}h_{22}-h_{12}^2}$$

where n, the number of random variables, is one less than the number of 1's in the symbol I. Moreover if $\tilde{t}$ and $\tilde{\phi}$ and are the true values, then

$$(t_0-\tilde{t})/s_t, \quad (\phi_0-\tilde{\phi})/s_\phi$$

follow the Student t-distribution with n-r degrees of freedom. We may thereby assign confidence limits to the estimates t_0 and ϕ_0, as obtained from the minimum value Q_0. If n-r were large, then t_0 would have a probability of 0.682 of lying within one standard deviation of the true value. We use the Student distribution in this way to find the equivalent range for a probability of 68%.

If σ is the standard deviation of individual deviations D_i, then $s^2 = Q_0/(n-r)$ is an unbiased estimate of σ^2. Moreover Q_0/σ^2 has a χ^2 distribution with n-r degrees of freedom, so that one may find limits of confidence for s^2/σ^2 at a given level, in our case chosen to be 95%.

There are canons, such as that known from Lalla's Śiṣyadhīvṛddhidatantra, in which the year cannot be effectively determined. In that case it is more practical to fix in advance the value of t, and to apply the present methods to determine ϕ_0 alone, in which case r = 1, in the above expressions.

4. DISCUSSION OF THE RESULTS

4.1 General remarks

For each of the canons, and each of the statistics I, the results are given in the table below, Section 5.

These results for the determination of the meridian are satisfying in that they lie in every case in India. Nevertheless, they differ to a varying extent from the meridian of Ujjain. There are two obvious ways in which this may occur, for we may have either an error on the part of the author of the canon, or it may have been the case that the observations were established for some other meridian, without any attempt to reduce them to that of Ujjain. No doubt both these reasons have some degree of application.

A third possibility is that the discrepancy might arise from the equation of time. I have indicated elsewhere, however, (Mercier 1985), that while difficulties would arise because of the way in which the equation is defined in Greek, Arabic and Latin usage, in Sanskrit usage the equation takes positive and negative values symmetrically, so that there the Mean Solar Time would be the same as that used now. This would seem to be the case from the beginning of Sanskrit astronomy, since Varāhamihira, in his discussion of the Sūryasiddhānta (Pañcasiddhāntikā IX,9) indicates an approximate rule for the equation.

4.2 The Romaka Siddhānta

The earliest of the canons in this survey is the Romaka

Siddhānta, based on parameters given by Varāhamihira in the Pañcasiddhāntikā (VIII, 1-8). The deviations are limited in use, since the lunar apogee and node are not in good agreement with the Sun and Moon. Therefore only the statistic (1 1100 00000) is available, and we can only find a year and meridian such that $Q_0 = 0$, since $n = r$. Therefore no statistical bounds can be assigned to the year and meridian. Nevertheless the results are very interesting, especially in regard to the meridian 86;40. This is considerably too large for one to suppose that Indian observations were responsible for the canon, but if we recollect the longitude difference of 56;30 between Alexandria and Ujjain, according to Ptolemy's Geography (IV,5,9; VII,1,63)

Alexandria	60:30
Ujjain	117;0,

we see that the Romaka canon might very well have been transferred to Ujjain, by means of such presumed positions, from a Greek source at a meridian around 30;0, which is indeed the correct longitude of Alexandria. The name Romaka, and the use of the year 365.24666..., indicate clearly enough the dependence on a Greek source. The longitude difference 60;0 between Yavanapura (Alexandria) and Ujjain is entailed by a remark concerning two astronomers of the time, Lāṭācārya and Siṃhācārya (Pañcasiddhāntikā, XV, 17-20) (2).

4.3 Āryabhaṭa

In the list of results in Section 5 one naturally groups together the next three canons, those of the Āryabhaṭīya, and the two from Varāhamihira's Pañcasiddhāntikā: the Sūrya Siddhānta, and the emendation to the planetary parameters of that work defined in XVI, 10-11. We know that only the Āryabhaṭīya is accurate in the case of Jupiter, so that D_8 must be omitted from the statistic for the other two. That much is clear from Billard's analysis of the deviations, and the use of the most recent modern ephemerides does not alter the position. The use of the new ephemerides has however an extremely interesting consequence, for the optimum year is now very nearly equal to 499 A.D., incomplete. That year is singled out by Āryabhaṭa (Āryabhaṭīya III,10) although he contents himself with merely telling us that he was then aged 23, so that one has never been altogether clear as to its astronomical significance. This point is however distant 3600 years, quite precisely, from the Kaliyuga, so that it can hardly be doubted that it had, for Indian astronomy, some decisive importance. When the older ephemerides of Newcomb, Brown, etc., are used to obtain the year and the meridian, we obtain

(0 1111 01111) 507.6 ± 4.05 79;8 ± 1;24.

It is now apparent that although with the new parameters, no great change results, nevertheless the results point more clearly than before to the year A.D. 499 as that year to which Āryabhaṭa referred observational results which were collected around that time. It is not clear why exactly that year was chosen, although it may have been believed by

Āryabhaṭa that the precessional correction vanished then. In any case, how can one avoid the conclusion that the Kaliyuga, 3600 years earlier was defined and fixed as a direct consequence? The two versions of that epoch, Sunrise and Midnight, are fixed exactly, so that in terms of the related canons, Āryabhaṭīya, and the Sūryasiddhānta, respectively, 3600 sidereal years separate the epoch from Noon 499 March 21. The enduring use of the Kaliyuga in all the following centuries testifies to the decisive importance of the work of Āryabhaṭa in the history of Indian astronomy (3).

The statistical results for the emendation to the Sūryasiddhānta show remarkably that when some later astronomers established it, the effect was to improve the accuracy at the epoch year A.D. 499. We now have 498.1 for the Sun and Moon alone, and 499.1 with the planets included. It is as if those responsible had access to the observational data, and were able to make better use of it so as to obtain even more accurate longitudes for the selected epoch year. Is it not wonderful in any case, that the most modern and accurate parameters help us to discover and appreciate better than ever before the very accurate Indian observations made nearly 1500 years ago?

4.4 The canon of Lalla

The canon taken from a set of emendations given to us by Lalla is singularly interesting, because unlike most others, indeed all others preceding it, it is in good agreement with observations over a long period. This means that one cannot determine the year by the method of least squares, although other considerations such as the form of the emendations, lead one to associate it with the late ninth century. The determination of the meridian, however, is secure, and indeed safer than in some other cases, precisely because the optimum meridian is so insensitive to the choice of year. One is struck in this case by the close approximation of the optimum meridian to that of Ujjain, reinforcing the view that the canon was the product of outstandingly careful work. In the diagram there are shown the graphs of σ against ϕ for the selection of statistics given in Section 5.

4.5 The Dṛggaṇita

This work was composed in Śaka 1353 (A.D. 1431-2), by Parameśvara, an astronomer of Kerala, and constitutes the exposition of the Dṛk ('Observational') system (Sarma (1963)). It is produced by a set of emendations applied to that version of the Sūryasiddhānta on which Parameśvara wrote his commentary (Shukla (1957)), or equivalently, to the Karaṇa Tilaka of Vijayanandin, ca. 950 (Rizvi (1963)). The emendations of the solar and lunar parameters were derived from eclipses of Sun and Moon observed by Parameśvara in the period 1398 - 1432 inclusive, from which he determined mean longitudes referred to the date with ahargaṇa 1651700 (essentially 4522 sidereal years), at Sunrise, which is A.D. 1421 March 29. The text in which the dates and circumstantial details of the eclipses are given has been edited by Sarma (1966). For that date he found the following: Sun 0,13;0, Moon 10,4;6, Apogee 3,9;57, Node 4,23;55. These figures are most interesting

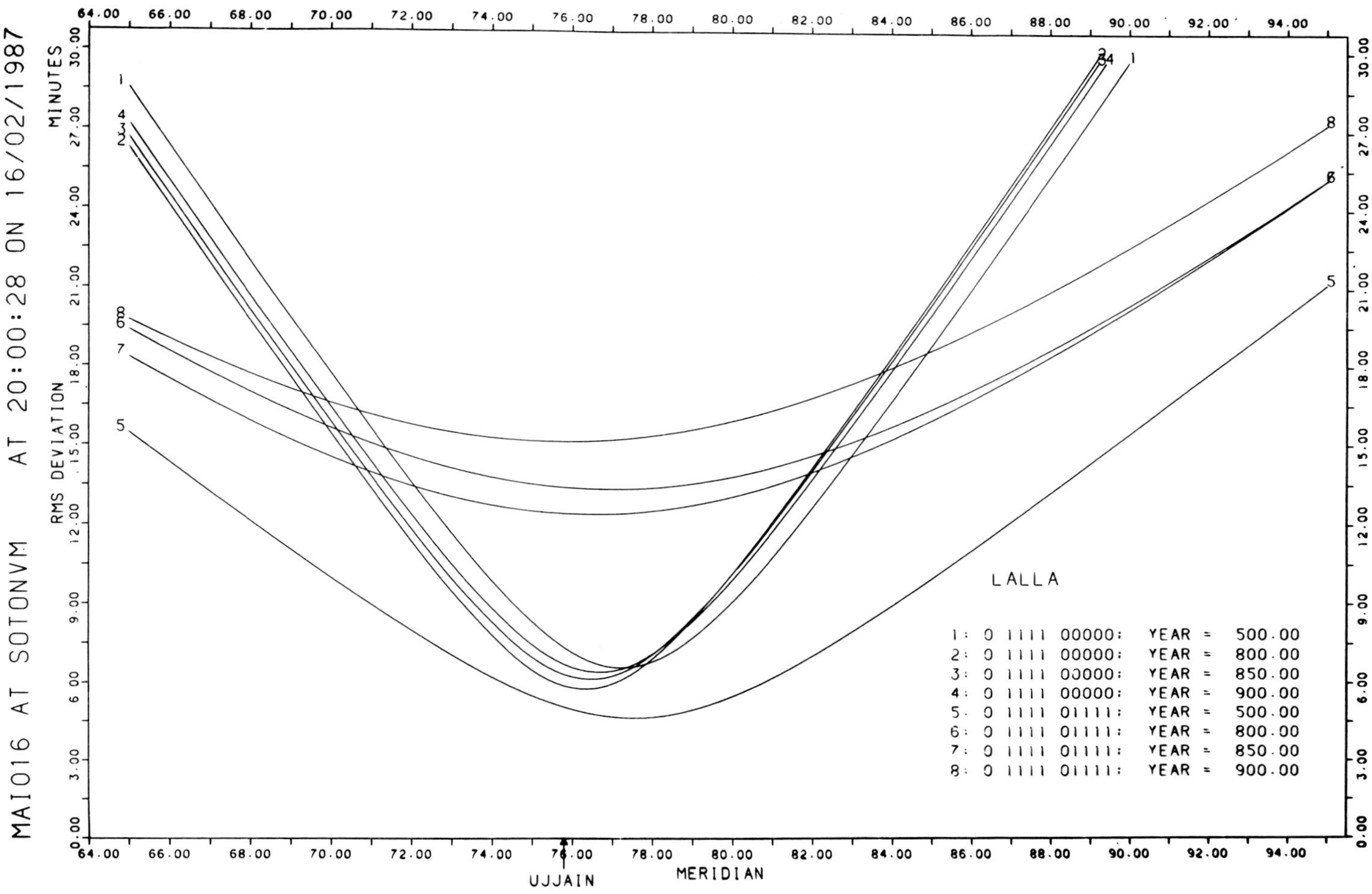

MAI016 AT SOTONVM AT 20:00:28 ON 16/02/1987
MINUTES
RMS DEVIATION
MERIDIAN
UJJAIN
LALLA
1: 0 1111 00000: YEAR = 500.00
2: 0 1111 00000: YEAR = 800.00
3: 0 1111 00000: YEAR = 850.00
4: 0 1111 00000: YEAR = 900.00
5: 0 1111 01111: YEAR = 500.00
6: 0 1111 01111: YEAR = 800.00
7: 0 1111 01111: YEAR = 850.00
8: 0 1111 01111: YEAR = 900.00

in that the longitude of the Sun is tropical, while the others are sidereal; that is the night-time observations are sidereal, the day-time tropical. Should we not see here an illustration of the paradigm of observation and tabular correction such as had been used throughout Indian astronomy, beginning with Āryabhaṭa?

The present methods applied to the eclipse deviations confirm well enough the documented circumstances, the year near 1421, and a meridian appropriate to Kerala (76;30) or perhaps Ujjain, for we cannot be certain whether Parameśvara reduced his parameters to the normal meridian.

4.6 Sphuṭanirṇayatantra

This work, by the Keralite astronomer Acyuta (1550-1620) was edited by Sarma (1974), and is of interest here because it is the one example known to me in Indian texts of a change of meridian. For, there is a short tract in another Malayalam MS, published as Appendix 5 by Sarma, in which the mean longitudes differ only by a shift of meridian, exactly 6.804 degrees Westward. Thus the respective meridians are approximately 80.5 and 74, neither of which however is suited to southern India. Moreover the accuracy is generally best in the thirteenth century, so one might infer that Acyuta obtained the canon from an earlier astronomer who lived further north.

5. TABLE OF RESULTS

The quantities given after the statistic are respectively, t_0, ϕ_0, with their standard deviations; σ, in minutes; the mean deviation at the point (t_0,ϕ_0); and the year and standard deviation for the meridian of Ujjain. In the case of those canons for which the year cannot be determined, the statistics concerning the meridian are given for certain preassigned values of the year. In all cases the year given is 'complete'.

Romaka Siddhānta
1 1100 00000

400.0	86;40			

Sūrya Siddhānta
0 1111 00000

498.1±31.67	77;19±6;25	10.67	0;14	493.8±14.7

0 1111 01101

502.0± 5.41	77;28±1;54	5.00	0;11	501.3±5.23

Āryabhaṭa
0 1111 00000

498.1±31.67	77;20±6;01	10.67	0;14	494.3±15.82

0 1111 01111

502.1±5.10	77;38±1;45	4.70	0;12	501.7±5.13

Pañcasiddhāntikā				
0 1111 00000				
498.1±31.67	77;19±6;25	10.67	0;14	493.8±14.7
0 1111 01101				
499.1±6.84	77;16±1.49	4.72	0;14	497.8±6.36
Lalla				
0 1111 00000				
500	77;11±2;40	6.6	0;5	
800	76;11±2;20	5.8	-4;47	
850	76;23±2;30	6.1	-5;35	
900	76;35±2;36	6.4	-6;24	
0 1111 01111				
500	77;23±1;43	4.6	0;8	
800	77;00±4;58	13.33	-4;44	
850	76;23±4;37	12.38	-5;34	
900	75;47±5;38	15.1	-6;24	
Karaṇa Tilaka				
0 1101 01011				
951.6±8.51	81;26±3;35	9.4	-6;59	950.7±10.43
1 1101 01011				
955.4±17.45	87;22±6;34	19.73	0;8	955.5±20.58
Dṛggaṇita				
0 1111 00000				
1424.5±27.19	77;19±6;6	9.9	-14;40	1420.5±12.30
Sphuṭanirṇayatantra				
0 1111 00110				
1250	80;47±1;45	4.64	-12;3	
1350	81;35±1;53	5.00	-13;39	
1550	81;47±5;39	14.99	-16;55	

6. SELECTED CANONS

There are three canons which are not given by Billard, and which are not readily available elsewhere.

6.1 Romaka Siddhānta

Epoch A.D. 505 March 21 Sunrise (1905588.75)

Sun	(150t-65)/54787 revs
Moon	(38100t-1984)/1040953 revs
Lunar argument	(110t+664)/3031 revs
Lunar node	-(24t+56278)/163111 revs

The time t is measured in days from the epoch. These results are taken from unpublished work on R. Billard.

6.2 Sphuṭanirṇayatantra (a)
Sphuṭanirṇaya tulyagrahamadhyamānayanam (b)

Epoch 588465.75 (Kaliyuga Sunrise)
Period (Kalpa) 1577917517019 days

	radix (a)	radix (b)	yugabhagaṇa
Sun	0;0	0;1,7,4	4320000000
Moon	4;21,21,36	4;36,18,7	57753321009
Apogee	118;50,9,36	118;50,17,11	488123229
Node	201;18,43,12	201;18,39,36*	-232297832
Mercury	350;12,28,48	350;17,7,14	17937072112
Venus	35;6	35;7,49,1	7022270775
Mars	348;9,21,36	348;9,57,15	2296862959
Jupiter	15;15,50,24	15;15,56,3	364172296
Saturn	340;22,48	340;22,50,17	146626695

*The text as edited gives 204;18,39,36, which must be emended to 201;18,39,36.

The radices (a) are equal to 0.4569 bhagaṇa, which recalls the construction used in the Brāhmasphuṭasiddhānta, which has 0.4567 bhagana as a general formula for its radices.

Both the texts (a) and (b) were edited by Sarma (1974), who gives (b) in Appendix 5.

6.3 Karaṇa Tilaka

The parameters are identical to those of the modern Sūrya Siddhānta in the version given by Billard, except for two yugabhagaṇa:

Lunar apogee	488211	in place of	488203
Lunar node	-232234	in place of	-232238

See Rizvi (1963).

NOTES

(1) The trigonometrical terms are available from the Bureau des Longitudes, Paris.

(2) I am indebted to some important unpublished work on this canon by Roger Billard, who proposed the interpretation of the meridian 86;40.

(3) This proposal, that the Kaliyuga originates strictly with Āryabhaṭa, contradicts van der Waerden's conclusions (1978 and 1980). His argument begins by observing that Abū Maᶜshar made use of the Kaliyuga as the date of the Deluge, and that he also made use of 'Persian' tables. If this only meant the Zīj-i Shāh in the mid-sixth century, there would be no problem, but van der Waerden argues for a Hellenistic dependence, via an earlier Persian system.

REFERENCES

Billard, R. (1971). L'Astronomie Indienne, Investigation des texts Sanskrits et des données numériques. Paris: École Française d'Extrême-Orient.

Francou, G., Bergeal, L., Chapront, J., and Morando, B. (1983). Nouvelles ephemerides du Soleil, de la Lune et des planetes. Astronomy and Astrophysics 128 124-139.

Mercier, R. (1985). Meridians of Reference in Pre-Copernican Tables. Vistas in Astronomy 28 23-7.

Rizvi, Sayyid Samad Husain (1963). A Unique and Unknown Book of al-Beruni, Ghurrat-uz-Zijat or Karana Tilakam. Islamic Culture (Haidarabad, India) 1963 (Apr., July, Oct.) 1964 (Jan., July) 1965 (Jan., Apr.).

Sarma, K.V. (1963). Dṛggaṇita of Parameśvara, critically edited with Introduction. Hoshiarpur, Punjab: VVRI.

Sarma, K.V. (1966). Grahaṇanyāyadīpikā of Parameśvara, critically edited with a translation. Hoshiarpur, Punjab: VVRI.

Sarma, K.V. (1974). Sphuṭanirṇaya-Tantra of Acyuta, with auto-commentary, critically edited with Introduction and ten appendices. Hoshiarpur, Punjab: VVRI.

Shukla, K.S. (1957). The Sūryasiddhānta edited with the commentary of Parameśvara. Lucknow: Lucknow University.

Spencer Jones, H. (1939). The rotations of the earth, and the secular accelerations of the Sun, Moon and planets. Mon. Not. R. astr. Soc. 99 541-558.

van der Waerden, B.L. (1961). Secular Terms and Fluctuations in the Motions of the Sun and Moon. Astronomical J. 66 138-147.

van der Waerden, B.L. (1967). Statistique Mathematique. Paris: Dunod.

van der Waerden, B.L. (1978). The Great Year in Greek, Persian and Hindu Astronomy. Archive for History of Exact Sciences 18 359-384.

van der Waerden, B.L. (1980). The Conjunction of 3102 B.C. Centaurus 24 117-131.

Armilary Sphere. It is an ancient China's cast-bronze astronomical instrument, engraved with excellent artistry. It has an observing tube, which is used mainly for determining the equatorial coordinates and the longitudes and horizontal coordinates of celestial bodies. It was made in 1437, the second year of the Zheng-Tong Reign of the Ming Dynasty.

3.5 PERIODIC NATURE OF COMETARY MOTIONS AS KNOWN TO INDIAN ASTRONOMERS BEFORE ELEVENTH CENTURY A.D.

S.D. Sharma
Physics Department, Punjabi University,
Patiala, India
India.

Apparitions of comets were thought to be a bad omen in earlier times in almost all the old civilizations. This led to correlating these apparitions with some particular events which took place simultaneously. Although the information was collected and recorded merely for astrological purposes, yet these records are in no way less important from astronomical points of view. Ancient Indian astronomers like Garga, Marīci, Asita, Devala and others made cometary studies and recorded their trajectories(Bṛhat-saṃhitā, Chapter on Ketucārādhyāya).

In earlier times there was a notion that the comets were heavenly bodies and their apparitions, paths, rising and setting in the sky, could not be found out by mathematical calculations as is clear from the following śloka(Bṛhat-saṃhitā, Chapter on Ketucārā dhyāya).

Darśanātamayo vā na sakyate jñātum /
Divyāntarikṣabhaumāstrividāḥ syaḥ ketavo yasmāt //

i.e. "The rising and setting times of comet(s) (*Ketus*) cannot be known by (mathematical) means because there are three types of *Ketus* (a more general name for some phenomena (celestial appearances) with names *Divya* (coming from sky), *Ānrtarikṣa* (appearing from places other than those of planets and constellations) and *Bhuma* (coming from the earth)". Although the algorithms for predictions of cometary positions could not be developed, yet the observed appearances, trajectories and physical characteristics etc. were recorded by ancient seers in all the old civilizations. Indian astronomers had the naked eye observations and compiled them in encyclopedic works known as *Saṃhitās*. It may be pointed out that although Varāhamihira (5th-6th century A.D.) in the opening of the chapter on comets (*"Ketu-Cāradhyāya"*) in his *Bṛhat-saṃhitā* expresses inability to predict the appearances and positions of comets, yet he has clearly stated the trajectory points among asterisms in case of some comets indicating that the motion was known to be periodic. For example, in case of *Calaketu* he gives the trajectory and other characteristics as follows:

Aparasyām calaketuḥ śikhayā yāmyāgrayāṅgulochritayā /
Gacched yatha yathodak tatha tatha dairghyamāyati //
Saptamunin saṃspṛśya dhruvamabhijitameva ca pratinivṛttaḥ /
Nabhasordhamātramittvā yāmyenastamupayāti //

i.e. "The *Calaketu* is one which rises in the west having tail directed towards south and about one *aṅgula* (one finger) in length, increasing in size as it moves towards north. It touches the Great-Bear, Pole Star and the Abhijit (Vega or α Lyrae) and then returns. Having described half part of the sky it sets in the south".

This shows that Varāhamihira was sure about the path of a comet. It may be noted that in the above-stated stanza the present tense is used in stating the various stages of the comet's apparition, which clearly indicates that this report is the final version of the universally factual phenomenon observed from time to time. It is in no way a single isolated report of the observation in the life-time of the author. The same trajectory is found reported in other saṃhitās too. This inference might be the result of investigations of earlier records and also Varāhamihira's own observations.

Another ancient astronomer, Bhadrabāhu (contemporary of Varāhamihira) in his Saṃhitā (named after his name as *Bhadra-bāhu-saṃhitā*) gives a categoric statement about periodic nature of motion of some short-period comet(s) as follows:-

> Sattriṃśat tasya-varṣāṇi pravāsaḥ parmaḥsmṛtaḥ /
> Maddhyamaḥ saptaviṃsastu jaghanyastu trayodaśa //

i.e. "the maximum period of disappearance of a comet is 36 years; the average period, 27 years; and the minimum period, 13 years".

This shows that Bhadrabāhu had knowledge about the periodic nature of cometary kinematics and it is believed that such reports could be verified in one's own life span too, by comparing the earlier records of apparitions and the physical characteristics (which help in recognizing the comet on its subsequent returns after disappearances).

Bhaṭṭotpala, (10th century A.D.) in his commentary on *Bṛhatsaṃhitā* (see Dikshit 1981) gives a list of time-periods of some long-period comets as follows:

> *Paitāmaha Ketu* has period of 500 years,
> *Auddālaka Śvetaketu* has period of 110 years,
> *Kāśyapa Śvetaketu* has period of 1500 years,
> *Vibhāvasuja Ketu* has period of 100 years, etc.

Note that here the names of comets are based on the names of the seers who estimated their time-periods. This tradition is similar to the one adopted in modern tradition of cometary studies.

From the above it appears that significant progress in cometary studies was made between the times of Varāhamihira (6th century) and Bhaṭṭotpala (10th century)

Edmond Halley in the year 1686 A.D. claimed periodic nature of motion of the comet, now known after his name as Halley comet and predicted his period to be about 76 years. He also determined periods and trajectories of some other comets using mathematical techniques based on the universal law of gravitation expounded by his contemporary Newton in his 'Principia'. This was a great advancement in the computational techniques for predicting positions of comets. As evidenced in this exposition, the Indian records show awareness of ancient astronomers with regard to the periodic nature of motions of comets. Some records on cometary kinematics are preserved in *Saṃhitās*. There are statistical data too, which deserve exhaustive mathematical analysis. The work in this direction is in progress and it will be interesting if some of the records of Halley's comet could be decoded from these texts.

REFERENCES

Bhadrabāhū-saṃhitā - Ed. Nemichandra Shastri, Bhāratiya Jnanapeetha, 1944.

Bṛhat-saṃhitā of Varāmihira(d.A.D. 587) - (1) Ed. with commentary of Bhaṭṭotpala by Sudhākara Dvivedi, 2 vols, Benares, 1895-97 (Vizianagram Skt. Series No.**10**); Reprinted Motilal Banarasidas (2) Ed. & Tr. V. Subrahmanya Sastri and M. Ramakrishna Bhat, 2 vols, Bangalore, 1947.

Dikshit,S.B.(1981). Bharatiya Jyotish Shastra, Pt.II, English Tr. R.V. Vaidya, p.343, Calcutta : India Meteorological Department.

DISCUSSION

O.P. Gingerich : There are about 100 currently known periodic comets, that is, comets that return in less than a few hundred years. Of these, only Halley's Comet is bright enough to be seen easily by naked eye. Hence, to discover the periodicity of comets with naked-eye observations, this comet provides the only possiblity. However, the period is somewhat variable, so that not until the last century were sufficiently precise perturbation calculation made to identify apparitions of Halley's Comet prior to 1531. Because the comet comes inside the earth's orbit, it is seen in quite different directions in successive appositions, and so there is no way to recognize the comet without specific computations of the elements of the ellipse. You may recall that Kepler analysed Halley's comet (1607) in terms of straight line motion in space. Thus, until you have a gravitational theory (conic sections and the law of areas) it is quite impossible to establish the periodicity of comets from observations alone. Hence I believe Prof. Sharma's conclusions are unwarranted and misleading.

S.D.Sharma : It is not to some extent possible to infer both periodicity and path of comets from naked eye observation. In fact physical characteristics (tail formation, colour etc.) may not be of much help, but the path can be ascertained certainly if conjunctions of comet with stars are recorded. In Bṛhat Samhitā, Calaketu is reported to occult Abhijit (α lyrae) and some other stars on its path. Such naked eye observation helps to ascertain the path among stars because the paralaxes of comets in general are not so large and node positions have quite less perturbation over a sufficient number of cycles of the comet. The periods might have been determined by noting velocity over the visibility period.

The occultations are meaningful in case of comets beyond 5 AU or even more. In case of Halley's comet, the difference in annual parallaxes in consecutive returns can be large, so it belongs to the category of "Aniyat-dik-prabhava" (Appearing in variable directional positions) comets as discussed in Bṛhat-Samhitā and it is not Son of any direction, instead it is Son of Brahma. (The God with four heads in all the four directions).

3.6 PLANETARY THEORIES IN SANSKRIT ASTRONOMICAL TEXTS

S.N. Sen
The Asiatic Society, 1 Park Street, Calcutta
India.

The origin and development of planetary theories in India are still imperfectly understood. It is generally believed that fullfledged planetary theories capable of predicting the true positions of the Sun, Moon and Star-planets appeared in India along with the emergence of the siddhāntic astronomical literature. Before this siddhāntic astronomy there had existed the *Vedāṅga Jyotiṣa* of Lagadha, prepared around circa 400 B.C. in the Sūtra period more or less on the basis of astronomical elements developed in the time of the Saṃhitās and the Brāhmaṇas. This *Jyotiṣa* propounded a luni-solar calendar based on a five-year period or *yuga* in which the Sun made 5 complete revolutions. Moreover, this quinquennial cycle contained 67 sidereal and 62 synodic revolutions of the Moon, 1830 sāvana or civil days, 1835 sidereal days, 1800 solar days and 1860 lunar days. An important feature of the *Jyotiṣa* is its concept of the lunar day or *tithi* which is a thirtieth part of the synodic month. The *tithi* concept was also used in Babylonian astronomy of the Seleucid period[1]. To trace the motion of the Sun and the Moon and to locate the positions of fullmoons and newmoons in the sky a stellar zodiac or a *nakṣatra* system coming down from the times of the Saṃhitās and the Brāhmaṇas was used. The *Jyotiṣa* was acquainted with the solstices and equinoxes, the variation in day-length of which a correct ratio was given. It is, however, silent about the inclination of the ecliptic, the non-uniform and irregular motion of the Sun and the Moon and various other important elements.

Of pre-siddhāntic astronomy mention should also be made of the Jaina Upāṅgas, the *Sūryaprajñapti* and the *Candraprajñapti*. These texts were written in ardhamāgadhī prākṛt possibly later than the *Vedāṅga Jyotiṣa* and taught a system of calendrical astronomy more or less identical with that of the *Jyotiṣa*. Then we have the *Gargasaṃhitā* (1st century A.D.) and *Paitāmaha-siddhānta* belonging to the same class and containing nothing new.

Āryabhaṭa's *Āryabhaṭīya*, compiled in A.D.499 and Varāhamihira's *Pañcasiddhāntikā* dated A.D. 505 were marked by a refreshing change in the treatment of the entire subject. The former is a succinct and highly condensed mathematical astronomical text developed around a unified sun-rise system, while the latter is a summary of five astronomical siddhāntas developed at different intervals of time and therefore of considerable historical significance. We know from Varāha that several astronomical siddhāntas other than the five he summarized in his book had been composed before his time, but these did not survive except in a very fragmentary manner in quotations and other references. It is quite clear that some time before his time Indian astronomy underwent a fundamental change and assumed a truly scientific, that is, mathematical, character. This change involved the following among others:-

(1) The archaic stellar zodiac was replaced by the twelve signs of zodiac;
(2) Besides the study of the Sun and the Moon, the study of five other star-planets and their motions was incorporated in an enlarged Yuga system;
(3) Sophisticated mathematics like continued fractions, rule of three, trigonometry-plane as well as spherical, indeterminate equations, and geometrical models were pressed into service to develop astronomical rules, formulas, and tables with the result that planetary motions could henceforward be explained mathematically and occcurence of eclipses be correctly computed before-hand.

These features characterized the compilation of astronomical siddhāntas during the next twelve hundred years until modern times, of which typical examples are Brahmagupta's *Brāhmasphuṭa-siddhānta* and *Khaṇḍakhādyaka*, modern *Sūrya-siddhānta*, Bhāskara II's *Siddhānta-śiromaṇi*, to mention a few, and several commentaries and Karaṇa works.

The planetary theories developed in these Sanskrit astronomical texts rest on three important concepts:-
(a) the Yuga, (b) the *ahargaṇa*, and (c) the geometrical models of eccentric circles and epicycles.

THE YUGA

The Yuga is a sufficiently long period of time in which the planets execute integral numbers of revolutions. Another property of this period is that the planets should find themselves at the fixed point of the Hindu sphere at the beginning of this period (and consequently also at the end of the same period). The *Vedāṅga Jyotiṣa*, as already mentioned, used a 5 year luni-solar period to accommodate the periodic motions of the Sun and the Moon. The *Romaka-siddhānta* as summarized by Varāha, used a luni-solar period of 2850 years. This period contained 1050 intercalary months and 16547 omitted lunar days. These elements reduce to the well-known Metonic Cycle of 19 years containing 7 intercalary months. The old *Sūryasiddhānta* summarized by Varāha had to use a still larger period, namely, 180,000 years in order to accommodate all the star-planets. Towards the end of the 5th and the beginning of the 6th century A.D. Āryabhaṭa adopted the Mahāyuga containing 4,320,000 years, which is exactly 24 times the period given in the old *Sūryasiddhānta*. Āryabhaṭa also adopted another smaller Yuga as one-tenth of his Mahāyuga, that is, a period containing 432,000 years which started on the midnight between February 17 and 18, 3102 B.C. He called it the Kaliyuga. Other Indian astronomers, Brahmagupta,

Bhāskara II and others, used a still larger period of time, the Kalpa which is 1000 times the Mahāyuga.

The origin of the concept of Mahāyuga for astronomical purposes is still an unsolved problem. In India, the Yuga system is met with in the laws of Manu and the *Mahābhārata*. The epic discusses the division of time and defines larger units of time like the 'Year of the Gods' (=360 ordinary year) and introduces the four mundane ages, e.g. Kṛta, Tretā, Dvāpara and Kali with the following lengths of time:

Kṛta Yuga	- 4800 years of Gods	=	1,728,000 years.
Tretā Yuga	- 3600 -do-	=	1,296,000 "
Dvāpara Yuga	- 2400 -do-	=	864,000 "
Kali Yuga	- 1200 -do-	=	432,000 "
	12,000		4,320,000 years

Similar ideas of great years are also met with among the Greeks, the Babylonians and the Persians. Early Greek philosophers like the Pythagoreans believed in the 'eternal return of all things' at the end of a long enough period of time. Furthermore, such cycles were marked by conflagration of great natural calamities. Van der Waerden (1980) who investigated the question of the origin of great years showed that the Greeks had a great year of Orpheus comprising 120,000 years and a great year of Cassandrus comprising 3,600,000. All these years are built out of factors 120 and 3600. The concept of a great year associated with flood and planetary conjunctions is often ascribed to the Babylonian priest Berossus who foretold that there would be a conflagration when all planets had a conjunction in Cancer and a deluge when such a conjunction took place in Capricorn. Berossus further estimated that the sum of regnal years of mythical kings before the Flood totalled 120 saroi or 432,000 years, 1 saros being 3600 years.

Did Āryabhaṭa adopt the period of 120 saroi or 432,000 years as his Kali Yuga period? It should be mentioned in passing that the number 432,000 can also be derived from more ancient Indian tradition such as 27 nakṣatras and the number of bricks 10,800 used in the construction of sacrificial fire-altars as mentioned in the *Śatapatha Brāhmaṇa*, for 4x27 = 108; 27x16 = 432; and 10,800x40 = 432,000. Whatever the source of Āryabhaṭa in adopting the span of the period he did not in all probability determine the starting point of the Kali Yuga at mid-night between February 17 and 18, 3102 B.C. by backward extrapolation of planetary positions from a certain date for the obvious reason that planets were not in conjunction on the mid-night of February 17 and 18, 3102 B.C. Van der Waerden (1980) has made an ingenious suggestion that Āryabhaṭa possibly got his clue from Persian tradition of observing and recording planetary conjunctions, particularly of the two slowest planets Saturn and Jupiter, and arrived at the date by a happy guess-work.

In this great cycle, the Mahāyuga or Kalpa, the revolution numbers executed by each planet as also by the apsides and nodes of the Moon are given leading to a fundamental table needed for all manner of planetary computations. Table I gives such revolution numbers for the Sun, Moon and five star-planets in accordance with the *Sūrya-siddhānta* as summarized by Varāha, Āryabhaṭa's *ārdharātrika* system as also reproduced in the *Khaṇḍakhādyaka* and Bhāskara I's *Mahābhāskarīya* and the modern *Sūrya-siddhānta*. The figures for inferior planets Mercury and Venus represent the revolution numbers of their conjunctions in respect of Sun, the *śīghra*, their sidereal revolutions being the same as that of the Sun. The table gives the number of civil days, or sāvana days reckoned from sun-rise to sun-rise in the Mahāyuga.

Table I - Planetary Revolutions in Mahāyuga

Planet	Sūrya-siddhānta summarised by Varāhamihira	Āryabhaṭa I (ārdharātrika system) reproduced in Khaṇḍakhādyaka and Mahābhāskarīya	Modern Sūrya-siddhānta
Sun	4320000	4320000	4320000
Moon	57753336	57753336	57753336
Mars	2296824	2296824	2296832
Jupiter	364220	364220	364220
Saturn	146564	146564	146568
Mercury	17937000	17937000	17937060
Venus	7022388	7022388	7022376
Moon's apogee	488219	488219	488203
Moon's node	232226	232226	232238
Number of civil days in a Mahāyuga	1577917800	1577917800	1577917828

It is to be noted that the starting point of these revolutions is the fixed point of the ecliptic, the first point of Aries and that planetary motions are uniform. From this table it is easy to determine the period of each planet in days and mean daily motion in hours, minutes, seconds etc (sexagesimal units). The mean position of each planet can also be readily computed if one knows the number of civil days elapsed from the beginning of the Mahāyuga up to the date in question. With certain modifications it can also be done from the beginning of any epoch such as the Śaka era.

The system has great simplicity and obviates the necessity of continuous observation. It is however necessary to observe each planet at intervals of time, note the deviations of observed positions from calculated ones and suitably amend the table of planetary revolution numbers to ensure perfect fit between observations and computations. These are called *bīja* corrections. Āryabhaṭa revised his own *ārdharātrika* tables in his own life time to produce his more accurate *audayika* system, and later astronomers did the same to keep astronomical tables in step with the reality of planetary motions.

AHARGAṆA

The number of civil or *sāvana* days elapsed from the beginning of a Mahāyuga or epoch necessary for computing the mean position of a planet on a given date is called the *ahargaṇa*. If the calendar were maintained in *sāvana* days according to the epoch, the *ahargaṇa* could be immediately found from such a calendar. Not so in a luni-solar calendar where dates are kept according to a *saura* year equal to the time taken by the Sun to travel through 12 signs of the zodiac, a lunar month equals the time interval between two newmoons, and a lunar day or *tithi* is equal to 1/30th of the lunar month. The computation of civil days involves the reckoning of intercalary months and omitted lunar days (*kṣayāha*). Most astronomical siddhāntas and Karanas give rules and procedures for *ahargaṇa* computations. The basic steps are as follows:-

$$t = y \text{ years} + m \text{ months} + d \text{ days}.$$

Let us suppose that, in this time period, the number of *saura*, lunar and intercalary months be represented by m_s, m_l and m_i, and the numbers of *saura*, lunar and omitted lunar days by d_s, d_l, and d_o. M_s, M_l, M_i, D_s, D_l, D_o represent similar elements for the time period of *Mahayuga*

Ahargaṇa a is given by

$$a = d_l - d_o \qquad (1)$$

Clearly $m_l = m_s + m_i$ (2a)

$$m_s = 12y + m \qquad (2b)$$

$$d_s = 30(12y + m) + d \qquad (2c)$$

d_l is variously given by

$$\left.\begin{aligned} d_l &= 30m_l + d \\ &= 30(m_s + m_i) + d \\ &= 30(12y + m + m_i) + d \\ &= [30(12y + m) + d] + 30m_i \\ &= d_s + 30\, m_i \end{aligned}\right\rbrace \qquad (3)$$

d_1 can be computed from any one form of eqn.3 provided m_i be known. m_i is calculated from any of the following relations:-

$$\left.\begin{array}{l} \dfrac{m_i}{d_s} = \dfrac{M_i}{D_s} \\[2ex] \dfrac{m_i}{m_s} = \dfrac{M_i}{M_s} \end{array}\right| \qquad (4)$$

With d_1 thus computed, d_o is determined from:

$$\frac{d_o}{d_1} = \frac{D_o}{D_1} \qquad (5)$$

The ratios M_i/D_s, M_i/M_s, D_o/D_1 are found from tables of planetary revolutions given in all Sanskrit astronomical texts. Values of M_s, M_1, M_i D_s, D_1, D_o as computed from a few texts are given in Table 2.

TABLE 2
REVOLUTION NUMBERS OF THE SUN, MOON AND ASTERISMS AS GIVEN IN VARIOUS TEXTS AND THE VALUES OF M_s, M_1, M_i D_s, D_1, D_o

No.of	Āryabhaṭīya, Mahābhāskarīya, Laghubhāskarīya	Khaṇḍakhādyaka, Sūryasiddhānta of PS	Later Sūrya, Soma, Brahmā etc. Siddhānta	Paitāmaha siddhānta of the Viṣṇudharmottara
revl. of the Sun	4320000	4320000	4320000	4320000
revl. of the Moon	57753336	57753336	57753336	57753300
revl. of the Asterisms	1582237500	1582237800	1582237828	1582236450
Civil days	1577917500	1577917800	1577917828	1577916450
Saura Months (Ms)	51840000	51840000	51840000	51840000
Lunar months (M1)	53433336	53433336	53433336	53433300
Intercal.months(Mi)	1593336	1593336	1593336	1593300
Saura days (Ds)	1555200000	1555200000	1555200000	1555200000
Lunar days (D1)	1603000080	1603000080	1603000080	1602999000
Omitted lunar days (D_o)	25082580	25082280	25082252	25082550

These ratios involve operations with large numbers and their simplification. Some of the simplified forms given in astronomical texts are given below:-

Pañcasiddhāntikā : $\frac{M_i}{M_s} = \frac{7}{288}$

Khaṇḍakhādyaka : $\frac{M_i}{D_s} = \frac{1}{976}\left(1 - \frac{1}{14945}\right)$

$$\frac{D_o}{D_1} = \frac{11}{703}\left(1 - \frac{1}{111573}\right)$$

The mean position of each planet is obtained by dividing the *ahargaṇa* by the period in days of the planet and converting the remainder in degrees, minutes and seconds. This is the mean position or *madhyama graha* because uniform speed of revolutions was assumed in constructing the tables. To account for non-uniform motions of planets and to introduce quantitatively the two corrections due to two inequalities called *manda* and *śīghra* the geometry of eccentric circles and epicycles was used.

GEOMETRY OF ECCENTRIC CIRCLES AND EPICYCLES FOR *MANDA* AND *ŚĪGHRA* CORRECTIONS.

For the first inequality, the *manda* correction, two eccentric circles of the same radius or one circle and an epicycle with its centre on the deferent has been used (Figs.1 and 2). A'AOP is the apse line or *nīcocca rekhā*

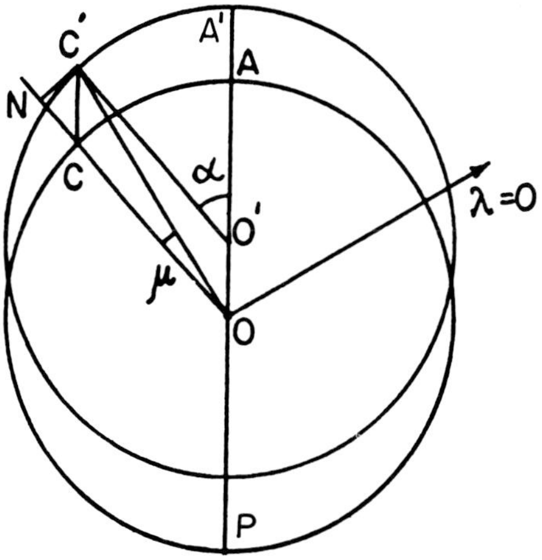

Fig-1
Eccentric model

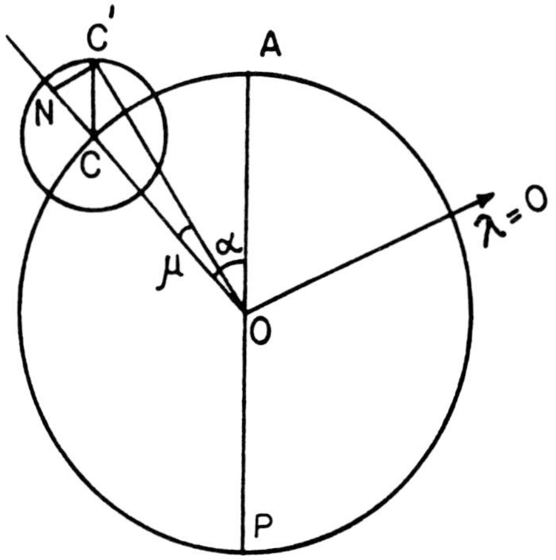

Fig-2
Epicyclic model

λ = 0 is the First Point of Aries or the starting point of planetary revolutions. The mean planet revolves on the deferent circle with centre O, while the true planet moves on the eccentric circle with centre O′, in the same anticlockwise direction with the same angular velocity. In the epicycle model the true planet moves on the epicycle with the same angular velocity but in the clockwise direction . The mean and the true planet both have the same longitude at the apogee (*mandocca*) and perigee, but in between the true planet lags behind the mean in one half of the circle and advances over it in the other half. The angular difference COC' is the equation of centre or *mandaphala* which may be expressed as follows:

$$R \sin \mu = \frac{R.r \sin \alpha}{\sqrt{(R + r \cos \alpha)^2 + (r \sin \alpha)^2}} \qquad (6)$$

where R is the radius of the deferent circle, r the eccentricity or radius of the epicycle, and α the anomaly, that is the difference of longitudes of mean planet and the apogee. If r be neglected compared to R, eqn.6 reduces to

$$R \sin \mu = \frac{R \sin \alpha \; 2\pi r}{2\pi R} = \frac{R \sin \alpha \; \odot_\mu}{360°} \qquad (7)$$

The astronomical siddhāntas prepared in the Āryabhaṭa school do not make this approximation in the hypotenuse for the *manda* correction while it is done in the Sūrya-siddhānta and other texts. $2\pi R$ or 360° is the circumference of the deferent circle or *Kakṣāvṛtta* and $2\pi r$ or $\odot_\mu$ the circumference of the epicycle.

THE SECOND INEQUALITY OR THE ŚĪGHRA CORRECTION

For the Sun and the Moon, the *manda* correction is sufficient, but for the remaining five star-planets, a further *śīghra* correction due to conjunction with the Sun is necessary. This is achieved by having recourse to another eccentric circle or an epicycle called the *śīghra parivṛtta*. The geometrical models are shown in Figs.3 and 4. After finding the corrected position C' upon *manda* correction, the *śīghra* epicycle is described with C' as centre. In it the true planet moves directly, that is anticlockwise direction, such that the direction of the planet from C' is always parallel to that of the Sun from O the centre of the deferent circle.

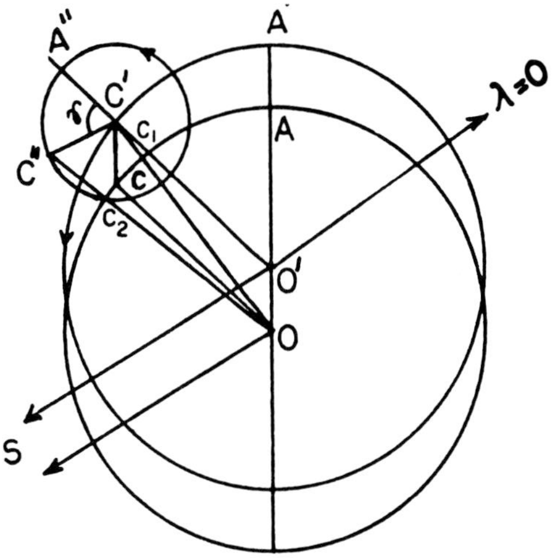

Fig -3

Eccentric - epicyclic model for planets other than the Sun or Moon

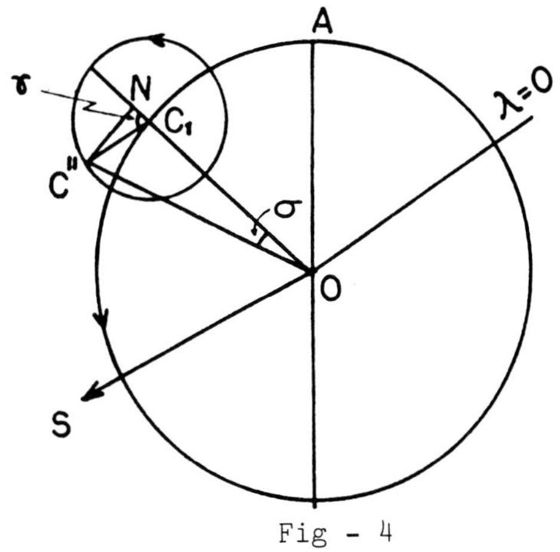

Fig - 4

Eccentric - epicyclic model for planets with approximation

Join C"O cutting the deferent circle at C_2 which is the corrected position of the planet. To simplify the calculations the *śīghra* epicycle is drawn with C_1 on the deferent circle as centre. C_1 is the intersection of C'O. The *śīghra* correction σ is given by

$$R \sin \sigma = \frac{R\, r_1 \sin \gamma}{\sqrt{(R + r_1 \cos \gamma)^2 + r_1 \sin \gamma)^2}} \qquad (8)$$

Here r_1 is the radius of the *śīghra* epicycle,γ the *śīghra* anomaly, that is, the longitude difference of the Sun and the mean planet after the *manda* correction.

It is clear from the formulae for *manda* and *śīghra* corrections that the values of certain constants must be known. These are the longitudes of apogee and the dimensions of *manda* and *śīghra* epicycles of different planets. These constants are given in all texts. An interesting feature of the epicyclic theory is that the dimensions of the epicycles vary according as the mean planets are in the odd or even quadrants.

Finally, the full values of the *manda* and *śīghra* corrections are not directly applied to the mean longitude of the planet to obtain its true longitude; instead a method of proceeding by half corrections in four steps is followed. These four steps given by Āryabhaṭa and explained by his commentator Parameśvara (Kālakriyā, 22-23) are as follows.

First step

The longitude of apogee (*mandocca*) is deducted from that of the mean planet and *mandaphala* is determined; half of the result is added to or subtracted from mean planet.

$$\alpha = \lambda - \lambda_A; \quad \mu_1 = \mu(\alpha); \quad \lambda_1 = \lambda \pm \mu_1/2$$

Second step

Deduct the longitude of the mean planet as corrected by step 1 from that of *śīghrocca* (Sun) and find the *śīghraphala*; half of the result is added to or subtracted from the mean planet corrected by step 1.

$$\gamma_1 = \lambda_S - \lambda_1; \quad \sigma_1 = \sigma(\gamma_1); \quad \lambda_2 = \lambda_1 \pm \sigma_1/2$$

Third step

The longitude of the apogee is deducted from that of the mean planet as corrected by steps 1 and II and the *mandaphala* is determined; the result is added to the original uncorrected mean planet to obtain the corrected mean planet (*sphuṭamadhya*),

$$\alpha_1 = \lambda_2 - \lambda_A; \quad \mu_2 = \mu(\alpha_1); \quad \lambda_3 = \lambda \pm \mu_2$$

Fourth step

The longitude of the (*sphuṭamadhya*) is deducted from that of the *śīghrocca* and the corresponding *śīghraphala* determined; apply the result to the (*sphuṭamadhya*) to get the true planet.

$$\gamma_2 = \lambda_S - \lambda_3; \quad \sigma_2 = \sigma(\gamma_2); \quad \bar{\lambda} = \lambda_3 \pm \sigma_2$$

A geometrical explanation of the procedure by half *mandaphala* and half *śīghraphala* was given by Neugebauer (1956). These steps are warranted by the simplifications and approximations introduced in the geometrical models and by the fact that the deferent circle is not large enough compared to the size of the epicycle.

Thus by the fifth century A.D. all the fundamental steps for computing planetary positions from a few set of tables were worked with fair degree of accuracy and embodied in Sanskrit astronomical texts. In the following centuries the methods continued to be elaborated and refined and formed the core of a sizable astronomical literature in texts, procedural manuals and commentaries.

NOTES

1 The concept of *tithi* is not attested in Greek texts. It is found in astronomical cuneiform texts, although, unlike the Sanskrit astronomical texts no name is given to the lunarday (Neugebauer 1952, 1955).

REFERENCES

Āryabhaṭīya, edited by H.Kern, 1874; English translation by W.E. Clark, 1930; edited by K.V. Sarma and K.S. Shukla with English translation by K.S. Shukla, (1976), Indian National Science Academy, New Delhi.

Bhattacharyya, S.P., and Sen, S.N., (1969), Ahargaṇa in Hindu astronomy, Indian Journal of History of Science, **4**, Nos. 1 and 2, 144- 155.

Candraprajñaptisūtra, edited with text in Hindi, Amolakrisi, Hyderabad, Vikramsamvat, 2445; see H.L. Kapadia, Indian Historical Quarterly, **8**, 1932 pp. 381-82; S.R. Das, Indian Historical Quarterly, **8**, 1932. pp. 36ff.

Neugebauer, O. (1952). The Exact Sciences in Antiquity, pp. 123, 178, Capenhagen.

-(1955). Astronomical Cuneiform Texts, **1**, p. 40.

-(1956), Transmission of planetary theories in ancient and medieval astronomy, Scripta Mathematica, **22**, 174, pp.

Pañcasiddhāntikā, edited by S. Dvivedi and G. Thibaut, 1889; edited by O. Neugebauer and D. Pingree, with English translation and notes, 1970-71.

Pañcasiddhāntikā, I, 15; the numbering of the verse is from notes, 1970-71.

Sūryaprajñapti, edited with the commentary of Malayagiri, Agamadaya Samiti, 1918. See G. Thibaut, Journal of the Asiatic Society of Bengal, **49**, 1880, pp. 107-127, 181-206.

Vedāṅga Jyotiṣa, with Somākara's commentary, edited by Sudhākara Dvivedi, 1908; also edited with his own Sanskrit commentary and English translation by R. Shamasastry, 1936.

Waerden, B.L. Van der (1980). The Conjunction of 3102 B.C.; Centaurus, **24**, 117-131.

DISCUSSION

S.M.R. Ansari : Are there various formulae for Ahargaṇa in Sanskrit texts, especially in later texts? Could you dilate on the basic difference between Pre-Ptolemaic & Indian Planetary model?

S.N. Sen : The ahargaṇa method given in astronomical siddhāntas remained the same in all texts, although different versions of the formulas, I explained, were used in different texts.

I have not made any special study of pre-Ptolemaic planetary theories and so I refrain from making any comparison.

S.D.Sharma : You have tried to explain that according to Ketakar śīghraphala converts heliocentric positions to geocentric ones. How? Why mandaphala and śigraphala are so named?

S.N. Sen : I have given my explanation according to astronomical siddhāntas. In the geocentric model, for planets other than the Sun and the Moon, a conjunction correction is needed. This correction is the śīgraphala and I explained how this is done in the eccentric-epicyclic model.

In the first inequality mandocca is the apogee and the anomaly is reckoned from the apogee. This anomaly is required for the calculation of the equation of centre and therefore it is called the mandaphala.

Likewise the śīghrocca is the farthest point in the śīghra epicycle when the superior planet is in conjunction with the sun. As the śīghra anamoly, that is, the longitude difference between Sun and the mean planet is required for the conjunction or śīghra correction, it is called the śīghraphala.

3.7 THE HSIU-YAO CHING AND ITS SANSKRIT SOURCES

M. Yano
Kyoto Sangyō University, Kamigamo, Kita-ku, Kyoto, Japan

1 *THE TEXT OF THE HSIU-YAO CHING*

The *Hsiu-yao Ching* (宿曜經 HYC) is a Chinese text on Indian astrology composed in the middle of the eighth century. Its full title can be rendered as 'Good and bad time and day and beneficient and maleficient mansions and planets promulgated by Bodhisattva-Mañjuśrī and other sages'. As the title shows the book is ascribed to the legendary Mañjuśrī and other sages, but the actual author is the Buddhist monk Amoghavajra (A.D.705-774) whose native place was somewhere in north India. His Chinese name Pu-k'ung Ching-kang is a literal translation of the Sanskrit name. Like most of the texts on Buddhist astrology and astronomy, HYC is contained in Vol.21 of the Taisho Tripitaka compiled by the Japanese Buddhist scholars during the Taisho Period (1912-1926). From many corruptions in the texts it seems that the compilers were not much interested in Buddhist astrology and astronomy in general, and that they did not try to secure better manuscripts either. Specifically in the case of HYC they simply based their edition on the text of the Korean Tripitaka and put in the footnotes the variant readings found in the Chinese Tripitaka of the Ming Dynasty.

There existed, however, several important manuscripts and even printed texts in Japan which escaped the attention of the Buddhist scholars of the Taisho Period. What I regard as the best text was published in 1736 by Kakusho, a learned Buddhist monk. He collated five manuscripts, in addition to the Chinese and Korean Tripitaka texts, and, as he says in his introduction, he based his text on a very old manuscript preserved in Muryojuin temple in Koya-san, one of the oldest sanctuaries of the Japanese Tantric Buddhism. He guessed that the manuscript was more than 500 years old. Actually the date is A.D.1160 according to Yamamoto (1975, p.28). Further Kakusho suspects that the manuscript could be directly traced to the one brought from China in 806 by Kukai, the founder of the Shingon school of Tantric Buddhism. When we compare Kakusho's critical edition with the Tripitaka text, we see immediately that Kakusho's is far superior. Recently I had an opportunity to examine another old manuscript of the same text possessed by Toji temple in Kyoto. The manuscript, copied in the 14th century from the one which was punctuated in 1121, was very close to Kakusho's edition. As a result of the study of this old Japanese recension, I could throw light on some important aspects of the origin and development of the text of HYC.

1.1 *Relation between the first and the second books of HYC*

In the both Tripitaka and Japanese recensions HYC consists of two books. According to the introduction of the first book, the text was first 'translated' by Amoghavajra in 759 and his Chinese disciple Shih-yao (史瑤) wrote it down. Since this first translation was a kind of a rough draft without proper arrangement of chapters, and since it was difficult for readers to use it, another Chinese scholar, Yang Ching-fêng (楊景風), revised it in 764 under the direct supervision of Amoghavajra. It is after this passage that the Kakusho edition adds an interesting remark:

> *Looking at the text there are two books: one is that which Shih-yao first wrote down, the other is the one which Yang Ching-fêng revised.*

Although this remark is an interpolation made in a later period, it gives us nonetheless important information, namely, that the two books are actually two versions of the same original text of Amoghavajra, and that the apparent differences are to be ascribed to his Chinese disciples. In fact almost all the topics of Book II (Shih-yao's first draft) are discussed in Book I (Yang's revision), the difference in fact being trivial. A reader of the Tripitaka recension, even after admitting the abundance of the subjects shared in both the books, might still argue that some topics discussed in Book II are not found in Book I or vice versa. For instance, the famous passage which provided a list of the names of days of week in Sogdian, Pahlavi, and Sanskrit is missing in Book I of the Tripitaka edition. In fact, however, Book I of the Kakusho edition adds these names in Chapter 4 where the influence of the seven luminaries is instructed. There are some other instances where the Tripitaka text cut off the original passages of the Book I, while the Japanese recension preserved them.

Fortunately, even after the completion of Yang's improved translation Shih-yao's first translation was not thrown away. This manner of keeping different kinds of translations, which is not uncommon in the case of Buddhist texts, caused a confusion on the part of Chinese readers and resulted in the frequent alternations of the text, as are found in the Chinese and Korean recensions, until finally the two translations were mistaken as the two consecutive books of one text. Japanese readers, on the other hand, who had no idea of mutilating a sacred Buddhist text, preserved both the translations almost intact. When we compare these two translations we can witness the process through which the original communication by the Indian monk was passed to the Chinese scriber and then to the reviser. One may call it a process of Chinization. The ascription of the new Buddhist text to Mañjuśrī is not strange, especially because Amoghavajra was the very person who intended to popularize the cult of Mañjuśrī worship in China. Now we can safely regard Amoghavajra as the author, instead of a mere translator, of the Hsiu-yao Ching. Of course it may be doubted whether he orally taught or he actually wrote down his work at all either in Sanskrit or in Chinese.

1.2 *Computation of the accumulated days (ahargaṇa) and the day of the week*

In the opening passage of the fourth chapter of Book I Yang

comments that the method of computing the seven planetary days is planned to be given in the end as the seventh ('eighth' in the Tripitaka text) chapter. Such a chapter is nowhere found in the Tripitaka recension, but it is preserved in the Kakusho edition as well as in the old manuscripts. The chapter is entitled 'Suan yao-chih'(算曜直 Computation of the day governed by the luminaries). The Chinese character 章 (chang) used here for 'chapter' is different from the one used for the other chapters 品 (p'in), because, as Yang comments, 'this is not what was taught by Mañjuśrī'. In fact the main body of this chapter is made of quotations from the *Chiu-chih li* (九執曆), which is a Chinese text on Indian astronomical calculations (karaṇa) composed in the second decade of the eighth century by Chut'an Hsita (瞿曇悉達 Gotamasiddha). This text was fully studied by Yabuuti (1979). According to a recent discovery (Zhāo 1978) Chut'an Hsita belonged to the third generation of the Indian family who worked for the T'ang Dynasty as official astronomers.

The procedure prescribed in the *Chiu-chih li* is essentially the same that is found in the *Pañcasiddhāntikā*, a compendium of the manuals of the five schools of Indian astronomy compiled by Varāhamihira in the sixth century. In order to find the day of the week one should first compute the number of the days (ahargaṇa) accumulated since the epoch. The algorithmic process of finding the ahargaṇa (D) can be expresed by the modern formulas:

$$M = (Y \times 12 + m) + \left[\frac{(Y \times 12 + m) \times 7 + k_1}{228}\right]$$

$$D = (M \times 30 + t) - \left[\frac{(M \times 30 + t) \times 11 + k_2}{703}\right]$$

where Y is the accumulated years since the epoch, m is the number of the elapsed months within the present year, M is the number of the accumulated month since the epoch, t is the number of days (actually tithi, which is 1/30 of a synodic month) elapsed in the current month, and k_1 and k_2 are epoch constants called kṣepa. Only the integer part of the result of the operation within the square brackets is taken.

The number of the accumulated days thus obtained is divided by 7 and the remainder tells us the day of the week. As Yang adds for the sake of his Chinese readers, one can obtain the number of the sexagesimal day-cycle if the division is made by 60 in the above process. Since the parameters in the formulas (7/228 for intercalation of months and 11/703 for omissions of tithi) are not precise enough, this rule works well only when the epoch is in the near past (i.e. when Y is small). Thus in the *Pañcasiddhāntikā* the epoch employed was Śaka 427 (=A.D.505).

Strangely enough the epoch used by Yang is 1501 years before the second year of K'ai-yuan (A.D.714). The epoch year thus is 788 B.C., which is the year of kuei-ch'ou (癸丑), the 50th in the sexagesimal year-cycle, as is stated in the text. To be more precise according to our text, the date of the epoch was the first day of the 'second' (Chinese) month (here Indian Caitra was intended), the day of chia-ch'en (甲辰) in the sexagesimal day cycle and it was also Monday. The day most probably corresponds with March 26 (Julian Days 1433691). Further Yang assumes that $k_1 = 0$ and $k_2 = 500$ without mentioning the reason. I wonder whether such a crude method was ever useful. Recently, however, when I visited the library in Koyasan University I was surprised to find a manuscript which was nothing but an evidence of a Japanese Buddhist who struggled with this computational method in as late as the 18th century.

The day of the week, or the day governed by one of the seven planet-gods in turn, is one of the chief subjects of HYC, the word yao in this very title meaning the '(seven) luminaries'. In Book II, namely. Shih-yao's original translation, we find an interesting passage which says

> *If it suddenly happnens that you do not remind of the day of the week, you are advised to ask it to the Sogdian, or Persian, or people from five Indias.*

Following this passage is the famous multilingual list of the names of planetary weekdays which will be given below. Probably Amoghavajra did not give any further information, but Yang Ching-fêng, when he prepared the revised translation, thought it necessary and useful to add the mathematical method of obtaining the day of the week, the method which he had acquired from the *Chiu-chih li*.

It is noteworthy that while this part of HYC which was added by Yang was soon cut off in China, it did survive in Japan for a long time. It seems that Kukai, the first of the three Japanese Buddhist monks who brought HYC to Japan, wanted to keep this chapter in order to compute the day of the week. I am not sure whether he ever attempted the computation. Concerning to this question, there is a very interesting folio which is added at the end of the Kakusho edition. Here we find the record of the first Sundays of the three consecutive years beginning with the 25th year of the Japanese era Enryaku, equated in the text with the first year of the Chinese era Yuan hê(元和), which began on 24 January, 806 . This is the very year when Kukai came back to Japan bringing HYC with him. In the colophone Kakusho says that the two oldest manuscripts in Koya-san contain this passage and he suspects that Kukai himself might have recorded the Sundays so that someone who did not understand the chapter on the 'Computation of the day governed by the luminaries' might be able to get a clue.

1.3 *The number of the lunar mansions, 27 or 28?*

In the Indian texts of astronomy and astrology the number of lunar mansions (nakṣatra) is sometimes 27 and sometimes 28. When 28 are counted Abhijit is put between Uttarāṣāḍha and Śravaṇa. It seems to me that when Amoghavajra first communicated the Buddhist version of Indian astrology to Shih-yao he intended only the 27 nakṣatra system. My conjecture is based on the fact that no mention of Niu (牛 , a word used for a Chinese translation of the nakṣatra Abhijit, but the star itself is not to be identified with the Chinese lunar mansion Niu) is found in Shih-yao's first translation except in one passage which is mentioned below. It is in Yang Ching-fêng's revision that one finds inconsistent references to Niu. In the first part of the first chapter Amoghavajra enumerates the correspondence between the 27 nakṣatra and the 12 zodiacal signs. In this case each nakṣatra occupies an equal space of 13;20° along the ecliptic. Thus one sign consists of two and one quater of nakṣatra. In the next paragraph Yang gives a list of the schematic allotment of a hsiu to each day. The list covers 12 months, each month consisting of 30 days. Here Kakusho's edition is based on the 27 nakṣatra system, which seems to be closer to the original, while the Tripitaka recension follows the 28 nakṣatra system. However, in the passage of Book I where the deity, gotra, etc. of each naksatra are summarized, the account of Niu (Abhijit) is added in both the recensions. In Kakusho's edition it is put at the end, while in the Tripitaka text it is placed between Tou (斗 Uttarāṣāḍha) and Nu (女 Śravaṇa). This is a typical instance of what I called Chinization. The original 27 nakṣatra system was modified into the 28 nakṣatra system so that the text might be more appealing to the Chinese readers. This Chinization explains well why the Tripitaka recension of HYC is so drastically different from the Japanese recension which has preserved the original reading with less alternation. The only disadvantage of the Japanese recension is that it did not make clear distinction between the main text and what was added by Yang Ching-fêng.

The custom of assigning the 27 nakṣatra to days was transmitted to Japan by HYC. Later in the seventeenth century the 27 hsiu system was replaced by the 28 hsiu system. It might be interesting for the modern reader to know that the old, and ultimately Indian, system of 27 nakṣatra is reviving recently in the popular Japanese almanacs.

2 *THE CONTENTS OF THE HSIU-YAO CHING*

Since Yang Ching-fêng's revised translation is better arranged, I follow his order of description and his naming of chapters in my brief sketch of HYC. Occasionally I refer to Shih-yao's original translation (Book II) which retains more of the Indian flavor. For reasons mentioned above I have used the Kakusho edition.

2.1 *The first chapter: On the Classification of Naksatra and Zodiacal Signs*

First of all the correspondence between the 27 nakṣatra and 12 zodiacal signs is briefly mentioned. The Sun and Moon have the equal share of 13 1/2 nakṣatra . Before the full description of the constituents of each sign and the planet-god presiding over it, the text gives the size (probably diameter) in yojana of the seven planets. A close parallel to this passage is found in the Buddhist text Abhidharmakośabhāṣya (III, 60) of Vasubandhu and in its Chinese translation. The values are quite different from those given in standard astronomical texts such as the Aryabhaṭīya I,5. The way of distributing the 27 nakṣatra to the 12 signs, in which the first point of Aśvinī is equated with that of Meṣa (Aries), is exactly that is prescribed in the Sanskrit astrological and astronomical texts which belong to the period after the Hellenistic influence. The description in HYC begins with Siṃha (Leo) which is governed by the Sun.

In the next passage the beginning of the Indian year is explained in the almost same wording that is attested in the Chiu-chih li (Yabuuti 1979, p.11). Here Yang explains the Indian way of naming the months according to the nakṣatra in which the full moon is stationed, as well as the correspondence between Indian and Chinese month names.

This passage is followed by the table in which the 27 nakṣatra are assigned to (12 x 30 =) 360 days as was mentioned above (1.3). At the end of this chapter the unequally spaced 27 nakṣatra are mentioned. The topic here is very close to that of the Śārdūlakarṇāvadāna (Mukhopadhyaya p.52) and the Pariśiṣṭa of the Atharvaveda I,5,6 (Bolling and Negelein p.4), the topic which belongs to the period before the Hellenistic influence on the Indian astrology.

2.2 *The second chapter: On the Lord of the Nakṣatra where the Moon stays*

For each nakṣatra the following items are recorded: the number of the stars comprising it, its configuration, deity, gotra, food, advisable and unadvisable actions, and general dispositions of the natives born under it. These are the standard subjects of the Indian astrological texts which can be classified as muhūrta-śāstra (Pingree 1980, p.101f.). In the last paragraph of this chapter the 28 nakṣatra are classified into the seven groups. The Chinese terms for these seven are translations of Sanskrit names dhruva, tīkṣṇa, ugra, laghu, mṛdu, mṛdutīkṣṇa, and carakarma, which are found in, for instance, Bṛhatsaṃhitā 97,6-10. In Indian texts there were some variants in the ways of classifying 28 or 27 nakṣatra into seven categories. Amoghavajra's version is closest to Parāśara's which is quoted in Utpala's commentary on the Bṛhatsaṃhitā. In the Śārdūlakarṇāvadāna similar terms are used (p.52 and 98), but the way of classification is quite different. It is only in this paragraph in Shih-yao's translation that the exceptional reference to Abhijit is attested.

2.3 *The third chapter: On the Secret Division of Nakṣatra into Three Nines*

The 27 nakṣatra are divided into three groups of nines beginning with the first, the tenth, and the nineteenth nakṣatra counted from the nakṣatra of one's birth in the order of increasing longitude. The Chinese names of these three nakṣatra are translations of janma, karma, and garbhādhānaka. Each one of them is followed by eight nakṣatra of which Chinese names are again translations of saṃpatkara, vipatkara, kṣema, pratyari, sādhaka, vaināśika, mitra, and atimitra, in this order. The idea propounded here is essentially the same that is briefly mentioned in Varāhamihira's Bṛhadyātrā 4,17 and it goes back to the Atharvanajyotiṣa which was quoted by Kane (1974, p.532f.). At the end of this chapter Amoghavajra explains the beneficient and maleficient influence of the seven luminaries over the native when they enter the nakṣatra thus classified. Commenting on this passage Yang says that the positions of the seven luminaries are to be computed by Indian method. After referring to three families of Indian astronomers, Gotama, Kāśyapa, and Kumāra, who were active in the T'ang Dynasty China, Yang says that he used Gotama's astronomical book. As we have seen above, the Gotama family were responsible for the transmission and establishment of the Chiu-chih li, the title of which Sanskrit equivalent is nava-graha-karaṇa (calculation of nine graha). Thus it is possible that the original text of the Chiu-chih li might have contained the section of the five planets and Rāhu and Ketu, besides that of the sun and moon which is available to us.

2.4 *The fourth chapter: On the Rule of the Seven Luminaries*

The seven luminaries, namely, Sun, Moon, Mars, Mercury, Jupiter, Venus, and Saturn, govern each day in turn in this order. This seven day week, which was first introduced in India by the Yavanajātaka in the second century, quickly gained a high popularity and became one of the most important elements of Indian astrology. It was transmitted to China in T'ang Dynasty, probably by Iranian people, especially by Manichaeans. In the Kakusho edition of Book I the foreign names are added at the beginning of each relevant paragraph prescribing the influence of the planets. As mentioned above Shih-yao (in Book II) provides a multilingual list of the names of the day of the week. It is to this part of Shih-yao's text that the two famous French Sinilogists paid a good attention (Chavannes & Pelliot 1913, p.171f.). Leaving out all the details of historical Chinese phonology I have simply tabulated the Chinese characters as they are used in Kakusho's edition of Book II and added the variant readings in Book I(K_1) and Taisho Tripitaka (T). The Sogdian and Indian (Sanskrit) names are those of the seven planet-gods, while so-called 'Persian' names stand for numerals 'one' to 'seven'. In the text each numeral is followed by 森勿 which is a phonetic translation of Pahlavi šambhih.

Table of names of weekdays

	Sogdian		'Persian'		Indian	(Sanskrit)
Sun	蜜	myr	曜	ēw	阿你底耶[1]	āditya
Moon	漠[2]	m'x	婁禍[3]	dō	蘓摩[4]	soma
Mars	雲漠[5]	wnx'n	勢	sē̌	盎峨羅迦	aṅgāraka
Mercury	咥	ṭyr	製[6]	čahār	部陀	budha
Jupiter	温没斯[7]	wrmzṭ	本	panǰ	勿哩訶婆跋底	bṛhaspati
Venus	那歇[8]	n'xyδ	數	šaš	戌羯羅[9]	śukra
Saturn	枳浣[10]	kyw'n	翕[11]	haft	賒及以室折羅[12]	śanaiścara

Variants 1. T 阿儞底耶 2. K_1 寞 T 莫 3. T 婁 4. K_1 蘇摩
5. K_1 雲漢 6. K_1& T 掣 7. K_1 鶻嗢勿 T 鶻勿 8. K_1 那頡
9. K_1 戌訖羅 10. K_1 枳緩 T 枳院 11. K_1 欲 12. K_1 拾室悉羅

2.5 *The fifth chapter: On the Miscellaneous Method of Secrecy*

Not all the five miscellaneous techniques of astrology instructed here are attestable in Sanskrit texts.

(1) According to the combination of the day of the planetary week and nakṣatra which the moon occupies, the days are divided into three categories. The Sanskrit equivalents of the Chinese names used would be Amṛta, Vajra, and Rākṣasa, although I can not give evidence.

(2) Divination according to the position of Venus. Thirty days of each month are divided into ten groups:(1,11,21), (2,12,22), and so on. To these ten groups are assigned the ten regions of Venus, namely, eight cardinal directions and the 'center' and the 'heaven'. I have been unable to find the Indian source of this peculiar idea.

(3) Six kinds of harmful nakṣatra. They are the 1st, 4th, 10th, 13th, 16th, and 20 nakṣatra counted from the one under which the native was born. The origin of this classification is also unknown to me.

(4) The houses and the planets. What is called 'life-sign'(命宮) here probably corresponds to the Greek hōroscópos or Sanskrit lagna, which is the sign on the eastern horizon at the time of the native's birth. Since, however, the text gave no definition, there was a room for several different interpretations among Japanese Buddhist astrologers. This is the only passage where we find a meagre reference to the twelve houses.

(5) A nakṣatra version of a zodiacal melothesia. If it happens that the client does not know the nakṣatra under which he was born, the astrologer can tell it by watching the part of the body which the client first touched when he appeared. The idea of a zodiacal melothesia of ultimately Egyptian origin was transmitted to India by the Yavanajātaka (Pingree 1978 II, p.199). Since, in the first chapter of HYC, the first point of Meṣa (Aries) is equated with that of Aśvinī, the latter should have been assigned to head. But Amoghavajra grafted the new idea of melothesia to the old nakṣatra system in which the first one was Kṛttikā, the third counted from Aśvinī.

2.6 *The sixth chapter on Śukla-pakṣa and Kṛṣṇa-pakṣa*

The Indian division of a month into two halves called śukla-pakṣa and kṛṣṇa-pakṣa is explained. For each of the 15 days (actually tithi) comprising a pakṣa, the name, deity presiding over it, and desirable and undesirable activities are instructed. A very similar idea with the closest terminology is expounded in Garga's work quoted by Utpala's commentary on Bṛhatsaṃhitā 98, 1-3.

2.7 *The seventh chapter: On the Computation of the Day Governed by the Luminaries*

As was already mentioned this chapter is not Amoghavajra's work but Yang Ching-fêng's addition of which a large part is made of quotation from the Chiu-chih li. The only, and most serious, difference lies in the epoch: the Chiu-chih li's epoch is the second year of K'ai-yuan (A.D.714), while Yang put it 1501 years back.

3 *CONCLUSION*

As we have seen above almost all the topics of HYC are attestable in Sanskrit texts on astrology and astronomy, although we can not say specifically from which source each topic was derived. It is natural that Amoghavajra was familiar with the Buddhist text Śārdūlakarṇāvadāna which was a part of the Divyāvadāna. This text represents the older period of Indian astrology when no evidence of the Hellenistic influence is found. At the same time he had some knowledge of new elements of astrology which were in vogue when he visited India, the elements which we now know from the works of, e.g., Varāhamihira. But Amoghavajra's knowledge of Indian astrology, as far as we judge from HYC, is far from professional. What few elements he added to the astrological instruction of the Śārdūlakarṇāvadāna were the equally spaced 27 nakṣatra and the seven-day planetary week. No mention is found to navāṃśa (1/9 of a sign), dvādaśāṃśa (1/12 of a sign), triṃśāṃśa (1/30 of a sign), dreṣkāṇa (1/3 of sign, decan), dṛṣṭi (aspect), etc. Probably he was not much informed of the genethliological astrology (Jātaka) of the Hellenistic origin in which the twelve houses (bhāva, Lat. domus) beginning with the ascendant (lagna) were the essential elements. Under the guise of the Buddhist 'sūtra'(經 ching), however, Amoghavajra's text survived among Buddhists, especially among Japanese Tantric Buddhists, and became the main source of the new school of Japanese astrology called Suku-yō-dō, Suku-yō being the Japanese pronounciation of Chinese Hsiu-yao, which again was the translation of Sanskrit Nakṣatra-graha.

It was only after the introduction of the Ch'i-yao jang-tsai tüeh (Formulas for preventing evil influence of the seven luminaries) and its ephemeride in 865 that Japanese Buddhists got informed of the horoscopic astrology and interested in obtaining planetary positions. Finally it was after the Fu-t'ien li was brought to Japan in 957 that they began to compute planetary positions in order to compose horoscopes (Momo 1975).

REFERENCES

Abhidharmakośabhāṣyam of Vasubandhu, ed. by P. Pradhan, Patna 1975.

Bolling, G.M. & Negelein, J. von. The Parśiṣṭa of the Atharvaveda, Vol.I Part 1, Leipzig 1909.

Bṛhadyātrā of Varāhamihira ed. by D. Pingree, Government of Tamil Nadu 1972.

Bṛhatsaṃhitā of Varāhamihira ed. by A.V. Tripāṭhī, Varanasi 1968.

Chavannes, Ed. & Pelliot, P. (1913). Un traité manichéen retrouvé en chine, Journal Asiatique, Janvier-Février 1913, pp.99-199.

Kane, P.V. (1974). History of Dharmaśāstra, Vol.V, Part 1, Poona.

Momo, H. (1975). Sukuyōdō to Sukuyōkanmon, *Rissho-shigaku* Vol.39, pp.1-20.

Morita, R. (1974). Mikkyo-senseiho, (reprint), Kyoto.

Neugebauer, O. & Pingree, D. (1971). The Pañcasiddhāntikā of Varāhamihira, Part II(translation & commentary), Copenhagen.

Pingree, D. (1978). The Yavanajātaka of Sphujidhvaja, Harvard Oriental Series 48. 2 vols.

Pingree, D. (1980). Jyotiḥśāstra, Wiesbaden.

Śārdūlakarṇāvadāna, ed. by Mukhopadhyāya, Śantiniketan 1954.

Yabuuti, K. (1979). Researches on the Chiu-chih li, *Acta Asiatica,* 36, pp.7-48.
Yamamoto, C. (1975). Hojuin no zōsho, *Mikkyogakkaiho,* No.14, p.11ff.
Yavanajātaka, see Pingree 1978.
Zhāo, H. (1978). A discovery of the Grave of Qū-tán Zhuan, *Wên-wu (文物)* 10, pp.49-53 (in Chinese).

DISCUSSION

L.C.Jain : Is there any mention on non-luminaries (dark planets) in the text Hsiu-Yao Ching (work composed in the middle of 8th Century) or any other text of Chinese astronomy, as there is a mention of certain dark-planets in Indian text ?

M.Yano : No, there is only mention of 7 grahas or luminaries, or else 9 planets alone. Two added are Rāhu and Ketu. Rāhu is same as in India. But there are many interpretations of Ketu.

3.8 THE POSITION OF THE *FUTIAN* CALENDAR ON THE HISTORY OF EAST-WEST INTERCOURSE OF ASTRONOMY

Shigeru Nakayama
University of Tokyo
Japan.

It is proved that the *Futian* calendar, a non-official one compiled in the Jianzhong reign period (780-783) in China, was brought to Japan in 957 by a Buddhist monk and was employed as the basis of horoscopes by the Buddhist school of astrology (Memo 1964). It was also used in competition with the official Chinese *xuanming* calendar for the usual functions demanded of a Chinese type lunisolar ephemerides, such as eclipse predictions. According to the view of the Song Dynasty Chinese scholar Wang Yinglin that the *Futian* calendar was "originally an Indian method of astronomical calculation" but Kiyosi Yabuuti has commented that Wang Yinglin's appraisal of the *Futian* calendar is solely based on a resemblance in form as it copied the trivial point of taking its epoch as the *Jiuzhi* calendar according to Indian astronomical methods and does not display a fundamental understanding of the Indian calendar (Yabuuti 1944).

We have lately discovered a fragmentary manuscript copy of the solar table of the *Futian* calendar, which was originally copied in 1230, transmitted by the Court Astrologer family with the title of *Futian lijing richancha licheng* and now preserved in the Tenri Library. I undertook to carry on the work of investigating the significance of the text from the viewpoint of the history of astronomy (Nakayama 1964). Its major findings were already summarized in English (Nakayama 1966) and here I present a fuller discussion of it.

ANALYSIS AND COMMENT

The table consists of the solar mean anomaly x in contrast to the equation of centre y in each Chinese degree. By analysing it, I found that this table is calcualted rather than observed as it precisely coincides with a parabolic function of

$$y = 1/3300 \; x \; (182-x)$$

though the maximum value of the solar equation of centre may have some empirical origin.

On the other hand, in Chinese official calendar in those days such as *xuanming* calendar the solar equation of centre is given in a round number in contrast to each 30^0 solar longitude. A simple glance of the table suggests us that these values given were possibly semi-empirical ones, by no means under a consistent mathematical treatment.

The Chinese smoothed over the variations in between these semi-empirical values by developing complicated interpolation formulae.

The Indian method, as seen in the *Jiuzhi* calendar, expresses it in terms of trigonometric function, the *Futian*'s parabolic treatment is not known in India (David Pingree's personal communication). It should be noted that the Indian calendar also employs Western degree of 360^0, which differs from the *Futian*'s traditional Chinese degree, *tu*, the number of one solar year. It should also be noted that the official Chinese calendrical scientists, who were unfamiliar with schematic and geometric representation and Indian trigonometric function, and only worked within the bound of their cherished algebraic tadition, must have approached a parabolic function, more advanced than a zigzag linear function, to such cases as parallactic elements in eclipse prediction (Nakayama 1969) as late as Ming dynasty.

The *Futian* calendar is undoubtedly advanced in mathematical treatment to other contemporary Chinese official calendars, while it cannot prove to be observationally superior. I show at the end of this paper a graph of solar equation of centre of several calendarical systems, in which apparently the *Futian* is not closer to modern value of Simon Newcomb. However, its unique mathematical feature was taken over by the later Uighur calendar and still further developed from *Futian*'s second degree to *Shoushi's* third degree algebraic function. In tracing this trend, I have an impression that the *Futian*'s mathematical formultion is a spontaneous development within Chinese tradition, rather than a product of Western influence.

In Chen Jiujin's recent article (Chen 1986) he argued that the *Futian* calendar was basically an importation from India under Buddhist influence, but as shown in the above, the close examination of the contents proved its independence of Indian influence. We can still safely say that the *Futian* calendar was practised among Buddhist sectors for the determination of their unique religious feast and horoscopic arts, as was the case in medieval Japan where the *Futian* calendar was employed by Buddhist monks in competition with the mainstream official Chinese calendrical system. It is still my unproven surmise that the different religious groups in China needed to adopt their own unique calendar for their religious observances and for that purpose, Confucian, Buddhist and Islamic calendrical offices and schools were maintained throughout Chinese history.

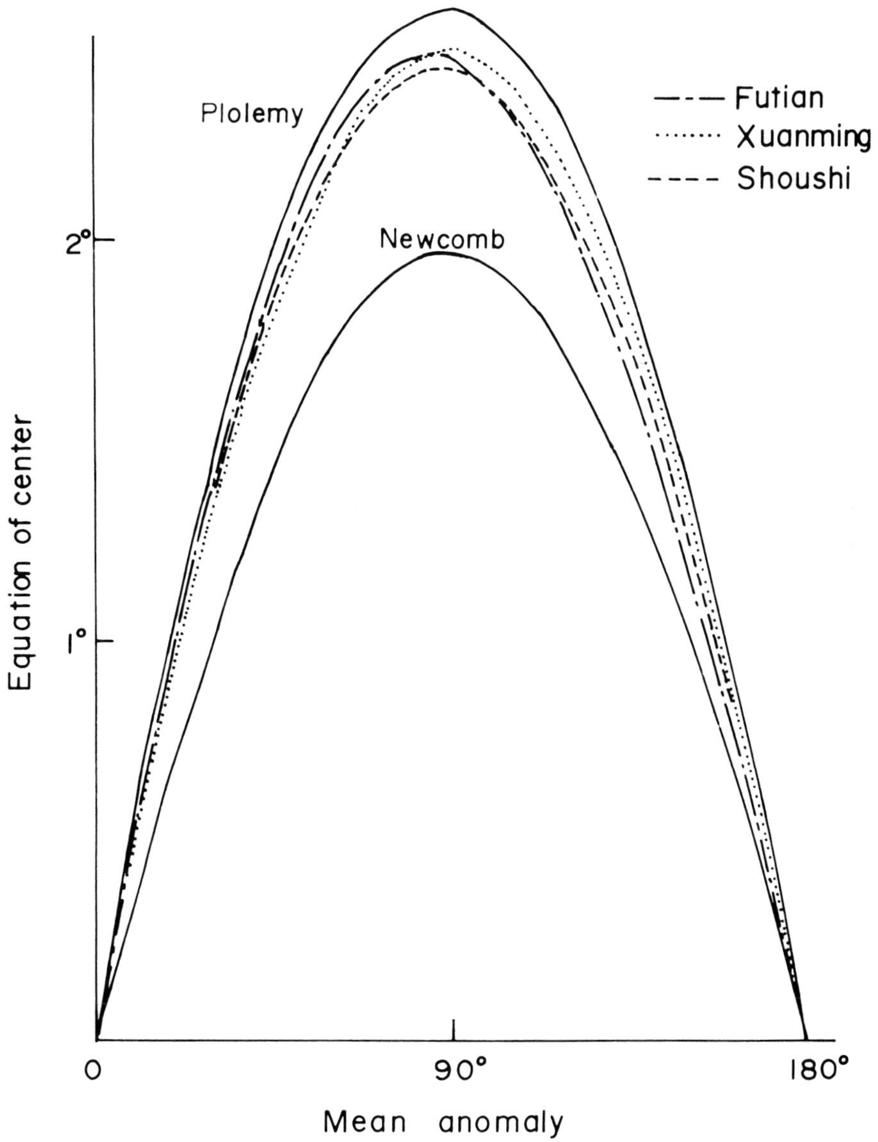

Fig - Solar equation of centre of several calendarical system

REFERENCES

Chen, Jiujin (1986)"A Study on the Futian Calendar". Studies in the history of natural sciences (in Chinese), **5**, no.1, 34-40.

Momo, Hiroyuki(1964). On the Futian Li, Journal of History of Science, Japan, **71**, 118-119(in Japanese)

Nakayama Shigeru(1964). The Significance of the Futian Li on the History of Astronomy, Journal of History of Science,Japan, **71**.120-122 (in Japanese)

Nakayama, Shigeru(1966). Characteristics of Chinese Astrology", ISIS, **57**,(4), No.90, 442-454.

Nakayama, Shigeru(1969). A history of Japanese astronomy, p.144, Cambridge,Mass. Harvard Univ. Press.

Yabuuti(1944). History of Sui and Tan Calendricl Science (In Japanese), 43-44, Tokyo, Sanseido.

DISCUSSION

L.C.Jain : Could you kindly comment on whether the Chinese parabolic function was related to spiro-elliptic function ? I have discussed these in Indian context in one of the papers on the spiro-elliptic orbit of the Sun in the Tiloyapannatti (Prakrit text of the 4th-5th Century A.D.).

S. Nakayama : The Chinese approach is purely algebraic rather than geometrical. Hence, it may be misleading to call it 'parabolic' but 'second degree equation'.

3.9 AN EXAMINATION OF AL-BĪRŪNĪ'S KNOWLEDGE OF INDIAN ASTRONOMY

M.S.Khan
Park Street Post, Box.No. 9072, Calcutta-700 016
India.

By the time al-Bīrūnī wrote the *KH* (see al-Bīrūnī's books; see also Boilot 1960; Kennedy 1970; Khan, A.S. 1982; Khan M.S. 1975) in the first half of the eleventh century, Greek astronomy specially the *Almagest* of Ptolemy was fully diffused among the Arabs. They also had some knowledge of Persian astronomy (Pingree 1963). But he was not the first Muslim scientist to write on Indian astronomy. Before he visited India (around 408/1017) he had some knowledge of this subject derived from Arabic translations of Sanskrit astronomical works. He records that the *Brāhmasphuṭa-siddhānta* of Brahmagupta (ca. A.D. 598) was translated by the order of Caliph Manṣūr of Baghdad (754-775 A.D.) in 154/771 or in 156/773(KH, p.351) and its Arabic translation was called *Sindhind al-Kabīr* or *the Great Sindhind*[1]. He adds that he translated it into Arabic and wrote a treatise on it[2]. He also studied its method of calculation in his treatise *Tarjumah mā fī Brāhma Siddhānd min Ṭuruq al-Ḥisāb* (Khan, A.S. 1982, p.20). Al-Bīrūnī states that he also retranslated the *Khaṇḍakhādyaka* by Brahmagupta, called in Arabic *al-Arkand* and published a new correct translation (KH, pp.346-383; Nalino 1911, p.83; Nadwi 1930, p.139). He had written a treatise entitled *Tahdhīb zīj al-Arkand* in which he corrected its errors but this is not available at present (Khan, A.S. 1982, No.6, p.12). The *Āryabhaṭīya* of *Āryabhaṭa* (b. A.D. 476) called *Arjabhadh* in Arabic was translated by Abu'l-Ḥasan al-Ahwāzī and it was perhaps available to al-Bīrūnī(KH, p.357; Pingree 1970, p.308; Nalino 1911, p.172-73).

About an important astronomical work, the *Siddhānta* of Pauliśa, al-Bīrūnī records that a translation of the whole work into Arabic had not been undertaken (*KH*, p.119; Pingree 1975, p.69-70). At another place he writes that he had started translating it but he could not finish it at the time of writing the *KH*(p.119, 315; Kern 1913, preface p.48).

Therefore, the parameters and computational techniques of the three schools of Indian astronomy were already known to al-Bīrūnī before his sojourn in India. He knew the elements and theories of Brahmagupta as expounded in his *Brāhmasphuṭa-siddhānta*. His *Khaṇḍakhādyaka* which followed the *Ārdharātrika* system of *Āryabhaṭa* was also known to him. Moreover, al-Bīrūnī also knew the system of *Āryabhaṭa* as explained in the *Āryabhaṭīya* and was well-versed in the Greek, Iranian and Arabic astronomical theories and methods.

When he visited India and came in direct contact with the Indian astronomers and Pundits, he had first-hand knowledge of other Indian astronomical works which he has mentioned in his *KH* such as the *Karaṇa Tilaka* (Rizvi 1963; Baloch 1973, p.94; Rizvi 1979) of Vijayanandī of Benaras based on the original *SS*, and the frangments of the original *Pauliśa-Siddhānta* and other works. He knew the commentaries of Balabhadra, *Bhaṭṭotpala*, Varāhamihira (ca. A.D. 505), Utpala, *Pṛthūdakasvāmī* (ca. A.D. 864) and others on the works of Brahmagupta and *Āryabhaṭa*. As regards the works of Varāhamihira, his *Bṛhatsaṃhitā* and *Laghujātaka*, the small book on nativities (astrology), were translated by al-Bīrūnī but they don't seem to be extant.

Al-Bīrūnī has discussed the Indian astronomical theories and parameters in his *KH* and other works also such as in his two *Rasā'il* in dealing with *Ẓilāl* (Shadows) and *Mamar* (*Transits*) (RIM, No.2, pp.1-226; RTM, No.3, p.107) published in 1367/1948. Some references to this subject are also found in his *QM*, *KAB*, *KHA*, *KTNA* and the *KT*. This paper will examine the knowledge of Indian astronomy as displayed by al-Bīrūnī with special reference to his *KH*.

After having explained that the *Siddhānta* is the general name given to every standard book on astronomy, he mentions the following five well-known *Siddhāntas*[3]

(1) *Sūrya-Siddhānta* i.e., the *Siddhānta* of the Sun, composed by *Laṭadeva* (505 A.D.)

(2) *Vaśiṣṭha-Siddhānta*, so called from one of the stars, of the Great Bear, composed by *Viṣṇucandra* (200 A.D.)

(3) *Romaka-Siddhānta* so called from Rome, composed by Śrīsena (about 600 A.D.)

(4) *Pauliśa-Siddhānta* so called after Pauliśa (Paulos), the Greek who composed it.(Sachau 1910,2, pp.304-305, note).

(5) *Brāhma-Siddhānta* so called from Brahman, composed by Brahmagupta.

Al-Bīrūnī states that he is in possession of the works of Pauliśa and Brahmagupta. But the table of contents of the *Brāhma- siddhānta* comprising twenty-four chapters as recorded by him does not fully agree with the arrangement of chapters found in its present editions (*KH* pp.119-20; see also Dvivedi 1902).

Al-Bīrūnī mentions the *Karaṇas* of *Āryabhaṭa* (?), Brahmagupta, Vijayanandī, Bhānuyasa and Utpala. Among the *Tantras* he refers to those of Āryabhaṭa, Balabhadra and Bhānuyasa. There are forty-six astronomers and astrologers mentioned by al-Bīrūnī in his *KH* (see also Sastri, A.M. 1975)

Al-Bīrūnī gives titles of eighteen *Purāṇas* (*KH*, pp. 100-101) of which he read parts of the *Matsya*, *Āditya*, *Vāyu* and *Viṣṇu Purāṇa* (*KH* p.101). The *Viṣṇu-dharma* or *Viṣṇudharmottara Purāṇa* was also well-known to him as he records the largest number of quotations from it concerning astronomical subjects and discusses them in detail. He takes Brahmagupta to task for accepting the statements of *Viṣṇu-Purāṇa* that the sun is lower than the moon (*KH* p.393). This is in connection with the eclipse of the Sun and the Moon and al-Bīrūnī criticizes these statements as unscientific and incorrect.

The question whether any book of *Āryabhaṭa* either in original or in Arabic translation was available to al-Bīrūnī or not may be asked here and an attempt may be made to answer it. He states in as many as three places in the *KH*, that he did not have access to any of the books of *Āryabhaṭa* adding that he knew his theories through the quotations given in the works of Brahmagupta and Balabhadra. According to the latest research, there are two *Āryabhaṭas* i.e., *Āryabhaṭa* I (b. A.D. 476) who is the same as the *Āryabhaṭa* of Kusumapura of al-Bīrūnī, and *Āryabhaṭa* II who flourished between 950 and 1100 A.D. It further complicates the issue as *Āryabhaṭa* of Kusumapura of al-Bīrūnī, is the younger and not the elder.[4] It can reasonably be suggested here that these statements of al-Bīrūnī refer to the astronomer whom he supposed to be *Āryabhaṭa* the elder. But in the case of *Āryabhaṭa* of Kusumapura he does not make such a statement and adds that he had read a book of *Āryabhaṭa* of Kusumapura (*KH*, p. 312). Thus he mentions three titles as three different works of *Āryabhaṭa* such as (1) *Daśagītikā*, *Daśakītak* in Arabic (2) *Āryaṣṭaśata*, Ar. *Ārjastasata* and *Tantra*, Ar. *Tantra* (*KH* p.121, 324). The first two are the titles of the two parts of *Āryabhaṭīya* of *Āryabhaṭa* (b. A.D. 476). It is this book which was translated by Abu'l *Ḥasan* al-Ahwāzī into Arabic (*KH* p.357; Nalino 1911, pp. 173-74).

Among Brahmagupta's works the *Brāhmasphuṭa-siddhānta*. and Balabhadra's commentaries on it were well-known to him. The *Khaṇḍakhādyaka* and *Uttara-Khaṇḍakhādyaka* were available to him in Arabic translations and used by him as has been stated in the introduction of a recently-published critical edition of *Khaṇḍakhādyaka* (Chatterjee 1970, 2, p.315).

The largest number of quotations concerning Indian astronomy found in the *KH* are from the *Brāhmasphuṭa-siddhānta*. of Brahmagupta and also from his *Khaṇḍakhādyaka*. Bina Chatterjee has compared the statements of al-Bīrūnī with the texts of the *Brāhmasphuṭa-siddhānta*. and *Khaṇḍakhādyaka* and found that by and large they are correct (Chatterjee 1975). She has pointed out a number of errors committed by al-Bīrūnī. The verses in the *Brāhmasphuṭa-siddhānta*. in the edition published by Sharma corresponding to the references of al-Bīrūnī to this text have been recorded by her (Chatterjee 1970, pp. 43-45).

According to al-Bīrūnī the *Khaṇḍakhādyaka* preserves the astronomical theories of *Āryabhaṭa's Ardharātrika* (Mid-night) system on which a commentary was written by Balabhadra. He writes about this book: "this Calander is the best known of all, and preferred by the astronomers to all others" (*KH* p.381). Bina Chatterjee, as already stated above, has made a comparative study of the quotations given by al-Bīrūnī with those found in the *Khaṇḍakhādyaka* (both *Uttara* and *Pūrva*) and this has generally speaking established the claims of al-Bīrūnī but this comparison is confined only to the *KH*.

Another astronomical work the *Pauliśa-Siddhānta* written in India was available to al-Bīrūnī[5] who had started its translation into Arabic but could not finish it by the time he wrote the *KH* (p.119). At another place, he states that the whole *Pauliśa-Siddhānta* "has not been translated into Arabic and gives reasons for this" (*KH* p.315).

The *Karaṇa-Tilaka* of Vijayanandī, which was translated by al-Bīrūnī into Arabic, has been mentioned in 4 of his works *RIM*, *RTM*, *KH* and *QM*. He has referred to it 10 times in his KH (Baloch 1973; Sastri, A.M. 1975, p.132; Rizvi 1963).

There are several other works of Indian astronomers and astrologers and their commentators such as Bhānuyaśa, *Bhaṭṭila*, *Kalyāṇavarman*, Durlabha, Parāśara, *Bhaṭṭotpala*, Śrīpāla, Utpala (considered as the same as *Bhaṭṭotpala*) and others who were known to al-Bīrūnī.

It is found that al-Bīrūnī had direct access to several genuine works of the outstanding astronomers of India and their commentaries. He derived the knowledge about the contents of other works through quotations found in these works and also perhaps from oral transmission. Balabhadra's commentary on the *Brāhmasphuṭa-siddhānta*. entitled *Bhāṣya* is lost but some quotations are preserved in al-Bīrūnī's works. None of the Arab or Indian Muslim astronomers who wrote on Indian astronomy after al-Bīrūnī could add anything new or substantial to the extent and the depth of the knowledge of this subject possessed by him. His knowledge about the Indian astronomy could not be more accurate and complete for various reasons. Most probably certain errors and discrepancies crept into the writings of al-Bīrūnī on account of faulty oral transmission by the Pundits.

It is evident that like a modern scientist, al-Bīrūnī follows a comparative method in his study of Indian astronomy and compares the astronomical methods and parameters of the Hindus, the Arabs, the Persians and the Greeks. He has also presented a comparative study of the 4 outstanding astronomers in India: Āryabhaṭa, Pauliśa, Varāhamihira and Brahmagupta dealing mainly with the circumference of the zodiac, the calculation of universal solar days, the Hindu tradition regarding the distances of the stars and the revolutions of planets in the *Caturyuga* and others.

Two modern scholars discuss how far al-Bīrūnī's references to the two basic works of Brahmagupta are correct. It has been stated that he was well aquainted with these two works. There are a few minor errors[6] which have been pointed out in them. The most frequent of these errors is that some of the statements found in the commentaries of *Brāhmasphuṭa-siddhānta*. and *Khaṇḍakhādyaka* have been attributed to Brahmagupta himself and full agreement is not always found between the Sanskrit texts and the statements of al-Bīrūnī. The occurrence of such minor error is quite possible in view of the fact that the commentaries contained both the original texts and their elucidations.

Al-Bīrūnī states that he never read any work by *Pṛthūdakasvāmī* yet he has ascribed to Brahmagupta some of the statements appearing in the former's commentary. At least two quotations from Brahmagupta found in the *KH* are not traceable at present. A statement of al-Bīrūnī giving rules for finding out the lords of a year and month which he claims to have copied from the *Khaṇḍakhādyaka* is not available in this book but found in its commentary by *Bhaṭṭotpala* in slightly altered form (Chatterjee 1975, p.165).

Al-Bīrūnī corrected the errors of the Arabic translations of Indian astronomical texts, studied them and offered his unbiased criticism on them. He also wrote original books on Hindu astronomy. The above study again confirms that al-Bīrūnī was a great synthesizer and transmitter of scientific knowledge[7].

The most important contribution of India to oriental astronomy are the planetary corrections which was discussed by Āryabhaṭa and the author of *Sūrya-Siddhānta*. But it is a mystery that al-Bīrūnī did not write about these planetary corrections. Perhaps it is due to the fact that the full text of the original *Pañcasiddhāntikā* of Varāhamihira was not available to him.

NOTES

1 It is generally believed that this book was the *Brāhma-Sphuṭasiddhānta* of *Brahmagupta* but al-Bīrūnī always mentions it as *Brahma Siddhanta* and incorrectly includes it among the five basic Siddhāntas. For the Indian delegation to the court of Caliph Mansūr see N.A. Baloch (1973, p.36), Pingree (1978,1975).

2 All references in this paper are to the text of the Hyderabad edition and its page numbers are given in the first bracket. Perhaps al-Bīrūnī's treatise on this book was entitled *Jawam'al-Maujūd li Khawāṭir al-Hunūd fī Ḥisāb at-Tanjīm*, D.J.Boilot(1960).

3 His explanation of the *Siddhānta* 'Straight, not crooked, nor changing' (*KH*,pp. 117-123) is incorrect. see the chapter XIV entitled *Fī Dhikr Kutubuhum fī Sā'ir al-'Ulūm* or on the Books of the Hindus on Scientific Subjects in the *KH*, pp.117-123.

4 *Āryabhaṭa* I of Kusumapura (modern Patna) born A.D.476 who wrote his *Āryabhaṭīya* in 499 A.D., while *Āryabhaṭa* II flourished between 950 and 1100 A.D. being the author of *Mahāsiddhānta*. See David Pingree (1970), Khan,M.S.(1978).

5 One of the works in which the citations from the original *Pauliśa-siddhānta* are preserved is the *KH* of al-Bīrūnī.

6 It has been proved by Bina Chatterjee on the basis of a comparison of the text of al-Bīrūnī's *KH* with those of the *BSS* and *KK*. See above. Cf. the exaggerated criticism of al-Bīrūnī's knowledge of Sanskrit astronomical texts by David Pingree (1968).

7 The scientific method which he followed in all his works specially in *KH* is the comparative method. The parameters and computational techniques of Indian astronomy have been compared by him with those of Greeks, Persians and Arabs.

REFERENCES

al-Bīrūni's books

1 Kitāb fī Taḥqīq mā li'l-Hind, Hyderabad ed. 1377/1958, pp.xliv + 30 + 548; Tr. English by E. Sachau, 2 vols. London, 1910; reprinted New Delhi, 1964 (Abbreviated as KH).

2 Kitāb al-Āthār al-Bāqiya 'Anil-Qurūn al-Khāliyah, ed. by Edward C.Sachau, Berlin, 1897 (Abbreviated as KAB).

3 Kitāb Taḥdīd Nihāyat al Amākin, ed. Md. Bin Tāwit al-Ṭanji, Ankara, 1962 (Abbreviated as KTNA).

4 Al-Qānūn al-Mas'ūdi, 3 vols, Hyderabad, 1954-56 (Abbreviated as QM).

5 Kitāb at-Tafhīm li-Awā'il Ṣinā'at-Tanjīm, Fasc. ed. and tr. English R.R.Wright as The Book of Instructions in the Elements of the Art of Astrology, London, 1934 (Abbreviated as KT).

6 Risāla Ifrād al-Maqāl fī Amr Az-Ẓilāl, Hyderabad, 1367/1948, pp.126 (Abbreviated as RIM).

7 Risāla Tamhīd al-Mustaqar li Taḥqīq Ma'na al-Mamar Hyderabad, 1367/1948, pp.107 (Abbreviated as RTM).

Baloch,N.A. (1973). Ghurrat az-Zījāt or Karaṇatilaka by Vijayanand of Benares, ed. and tr. Sind, Pakistan, Instruction.

Boilot,D.J. (1960). al-Bīrūnī in Encyclopaedia Islam, New Edition, **1**, 1236-1238.

Chatterjee,Bina (1970). Khaṇḍakhādyaka of Brāhmagupta, critically ed. with commentary of Bhaṭṭotpala and English translation, 2 vols, World Press, Calcutta.

Chatterjee,Bina (1975). al-Bīrūnī and Brahmagupta, Indian Journal of History of Science, **10**,161-165.

Chatterjee,Bina (1977). Śiṣyadhivṛddhidatantra of Lalla,ed. with comm. of Mallikārjuna Sūri (A.D. 1178) and English tr. Indian National Science Academy, New Delhi.

Dvivedi,S. (1902). Brāhmasphuta - Siddhānta of Bramagupta, ed. with his own comments, Benares.

Hall,F. (1859). Sūryasiddhānta, ed. with commentary of Ranganātha BI, Asiatic Society, Calcutta.

Kern,H. (1865). Bṛhatsaṃhitā of Varāhamihira, Calcutta.
Kennedy,E.S. (1970). al-Bīrūnī in The Dictionary of Scientific Biography, **2**, 147-158, New York.
Khan,A.S. (1982). A Bibliography of the Works of Abu'l-Raihān al-Bīrūni, Indian National Science Academy,New Delhi.
Khan,M.S. (1975). A Select Bibliography of Soviet Publications on al-Bīrūnī, Janus, 279-288.
Khan,M.S. (1978). Aryabhata I and al-Bīrūnī, Indian Journal of History of Science, **12**, 237-244.
Nadwi, Syed Sulayman (1930). Arab Wa Hind Ke Ta'lluqat, Allahabad, p.139.
Nalino, Carlo (1911). 'Ilm al-Falak Ta'rīkhuhū 'ind al-'Arab fī al-Qurūn al-Wusṭa, Rome.
Pingree,D. (1963). Astronomy and Astrology in India and Iran, ISIS, **54**, 229-246.
Pingree,D. (1968). Fragments of the works of Ya'qūb ibn Tāriq Journal of the Near Eastern Studies, **27**, 97-125; also see vol.29.
Pingree,D. (1970). Aryabhata I, Dictionary of Scientific Biographies, **1**, 308-310.
Pingree,D. (1975). al-Bīrūnī's Knowledge of Sanskrit Astronomical Texts in The Scholar and the Saint, ed. Peter J. Chwolkowski pp. 67-8, New York.
Rizvi,S.S.H. (1963). A unique and unknown book of al-Bīrūnī's Ghurrat az-Zījāt or Karana Tilaka, Islamic Culture, Hyderabād.
Rizvi,S.S.H. (1979). A Newly Discovered Book of al-Bīrūnī: Ghurrat az-Zījāt and al-Bīrūnī's Measurements of Earth's Dimensions in al-Bīrūnī commemorative Volume, ed.Hakim Mohammed Said, Karachi, pp.844, 605-680).
Sachau,E.(1960). al-Bīrūnī's India, Eng. tr. of KH, 2 vols, London, Reprinted, New Delhi, 1964.
Sastri,Ajoy Mitra (1975). al-Bīrūnī and his Sanskrit Sources, Indian Journal of History of Science, **10**, 121-123.
Thibaut,G & Dvivedi,S. (1889). Pañcasiddhāntikā of Varāhamira, ed. with Original Comments in Sanskrit and English translation & Introduction, Bombay.
Viṣṇudharmottara, Purāna, ed. Priyabal Shah, Gaekwad Oriental Series No.**130**, Baroda, 1958.

Gnomon. This is one of China's most ancient astronomical instruments. The meridian horizontal plate with the scale on it is called qui and the higher vertical pillar is called biao. The number of days in a tropical year and the 24 divisions of the solar year can be determined by the shadow of the biao at noon. It was made in 1439, the 4th year of the Zheng-Tong Reign of the Ming Dynasty.

3.10 THE ANCIENTS' CRITERION OF EARLIEST VISIBILITY OF THE LUNAR CRESCENT: HOW GOOD IS IT?

M. Ilyas
School of Physics, Universitiy of Science of Malaysia, Penang

Abstract. Earliest visibility of the lunar crescent is an important calendrical element. It was needed in all early calendars and remains in use in some lunar calendars today. An astronomical criterion of earliest lunar visibility was therefore evolved quite early, using observations, right from the Babylonian era. In subsequent periods the Babylonian single factor 'moonset lag' criterion was used extensively, although gradually it was realized that it was rather simple. Recently, an improved and comprehensive global criterion of earliest visibility, developed by the author, has been used to generate an extensive inverted moonset lag data set. These data, as a function of latitude and season, all for the first time provide a useful comparison with the simple ancient criterion. It is found that the simple criterion is remarkably good for the latitude region where the Ancients collected their observational data, illustrating the care with which their data was gathered. At other latitudes, there are significant differences, as may be expected. Although the simple criterion may now be replaced by an accurate season and latitude dependent criterion, the former will continue to provide a useful basis. The paper discusses various related historical developments.

Introduction

The easily observable monthly lunar cycle and the rapidly changing lunar phases make the moon an obvious choice for an unaided, simple, yet accurate natural timekeeping system. It is not surprising then that lunar calendrical practice is very old. Almost all early civilizations, Babylonian, Aztec, Inca, Hindu, Chinese, Greek, Jewish, and Muslim, made use of the lunar system in the past, as many do today. And if it was not for the Caesar's Pontiff playing around with the intercalation practice, and then for the Christian Church's need to tackle the problem of Easter date (McNally 1983), perhaps most of us, including the West, still would have been using the lunar calendrical system in either a pure or mixed (i.e. luni-solar) form as do the Muslims, Hindus, Jews and Chinese today.

Determination of when to expect the new lunar crescent's first visibility was a primary scientific challenge for the Ancients (it still is!). However, the Babylonians, quite early, had established a simple one-parameter rule of 'moonset lag' to determine the start of a new month. It remained in calendrical use with the later communities of the Middle East, Hindus, Greeks, Chinese, and early Muslims -- without any significant change or improvement. It was only around 500 A.D. that the Hindus began to recognise the importance of other parameters, especially the lunar crescent's width, in the determination of the earliest visibility. Later, Muslim astronomers like al-Battānī, al-Khwārizmī, al-Farghānī, and Ḥabash studied the problem more thoroughly and provided a more comprehensive and universal solution. Although al-Battānī is reported to have remarked: "the Ancients did not understand the phenomenon completely, but only approximately", he nevertheless appreciated its usefulness by stating that the simple criterion was a "good starting point" (Bruin 1977). But how good?

Development of a modern comprehensive criterion of earliest lunar visibility was first attempted in the early years of this century by Fotheringham and Maunder who made use of Schmidt's twenty year long careful observations of youngest lunar crescents at Athens. The information was inverted into a two-parameter criterion that could be considered of universal nature. But it remained obscure for the next sixty years, when the current Islamic interest brought it to the surface. Around this time, an independent theoretical treatment of the problem by Bruin (1977) was presented. Subsequently, the two criteria were combined (Ilyas 1981), and the composite criterion was further improved leading to an updated, observational/theoretical compatible prediction system with greater confidence (Ilyas 1983; 1984a; 1984b). The new criterion opened the way for the necessary global lunar visibility calculations and International Lunar Date Line work through a computer-based global calculation system (Ilyas 1982).

Since the Babylonian criterion was in use for a remarkably long period, and included the Muslim astronomers of later centuries like al-Ṣūfī and al-Kāshānī among its users, it is of great interest to examine the 'moonset lag' as a universal criterion for lunar visibility. The global calculation system enabled us to develop a new 'moonset lag' criterion and thus allowed a test of validity for the Ancient's simple criterion on a global scale. It is the purpose of this paper to discuss the results of this analysis.

New Moonset Lag Criterion

The Global Lunar Visibility Prediction Calculation System (or global calculation system) is a self-contained extensive algorithm which has been discussed in detail elsewhere (Ilyas 1984b). Briefly, it incorporates a number of basic solar and lunar positional programs, and enables the generation of various parameters, including moonset and sunset times at any specified (or grid-point) locations. For each new moon, a longitude (λ_o) is identified at which the minimum condition of visibility at local sunset has been met, latitude by latitude. At these λ_o various parameters are calculated and printed out. The moon's age and moonset lag at these critical longitudes (λ_o) provide the critical data to construct a simple, one-parameter criterion of lunar visibility. The results, based on sixty-one lunar months (1979-1983) at different latitudes, are shown in Fig. 1. The 'moonset lag' data at different (northern) latitudes (0°, 30°, 40°, 50°, 60°,) are plotted as a function of season (day-number of year: 1-365). We notice that there is a slight seasonal trend at lower latitudes, but the data (not shown here) are essentially confined to the curves. However, at higher latitudes, not only is the seasonal trend much more strong, but there is also a large scattering of the data, with the effect being particularly strong in the summer season. This is shown by the envelopes or dotted lines around the mean lines — (the seasonal dependence prevails at the southern latitudes (Ilyas 1984b)). At the lower latitudes, the data may be reasonably defined by lines of constant 'moonset lag' (minutes), but at the higher latitudes the departure is too significant:

0° : 41 ± 1 (±2 all data)

30^0 : 46 ± 2 (±4 all data)
40^0 : 49 ± 4 (±9 all data)
50^0 : 55 ± 10 (±15 all data).

Discussion

The results in Fig. 1 enable us to compare, in a straightforward manner, the Ancients' criterion of earliest lunar visibility, moonset lag greater than 48 minutes ($a_s \geqslant 12^0$). Considering the latitude of the place of observation, Babylon (32^0N), we notice that although strictly speaking, for the comparable latitude, the criterion deviates from this single factor —seasonally independent— criterion, the difference is rather insignificant. The fact that seasonal trend is not strong at these latitudes must have been noted by these early observers, thereby settling for a single factor criterion. Of course, we can quickly notice that there is a strong latitude dependence in the criterion, and therefore at the higher as well as the lower latitudes there are significant departures. At the lower latitudes, although the seasonal effect is almost negligible, the absolute magnitude of the Ancient's criterion is an overestimate.

On the whole, the Babylonian criterion is an excellent estimation, applicable to most of the lower latitudes. This must have been why it remained in favour with most of the later astronomers, including the Hindus and Muslims, who were located at the lower latitudes. Still, as al-Battānī pointed out, it was an excellent but approximate estimate. It is not clear to me whether the Babylonians arrived at this figure of "48 minutes" from the "age" of earliest visibility, knowing that approximately the moon had to be about "one day" old to become visible. A simple treatment shows that on the average the moon elongates from the sun by slightly more than 12^0 per day, and as a result the moon transits a meridian 49 minutes later than the sun on each subsequent day (see Appendix 1). However, it seems to be more likely that they would have directly noted the separations (in degrees) of youngest visible crescents over a long period and arrived at the excellent figure of 12^0 separation. This is because of the fact that to produce the same separation, the moon's age varies more significantly, giving a wider spread (Fig. 2). In any case, the Babylonians must have collected their observational data with great care, which produced the remarkably good simple criterion.

Concluding Remarks

We notice that the Babylonian simple criterion is a good starting point up to mid-latitudes. Of course, for higher precision one should employ the season-latitude dependent curves even at the lower latitudes. Even so, the 'moonset lag' criterion remains quite simple considering the accuracy. The 'visibility-criterion proper' is somewhat involved and requires considerable computation. But for chronological purposes of earlier records, one can relatively easily make use of the moonset lag criterion with considerable confidence. Also, to a layman this criterion is more meaningful, since he can easily understand that the (local) moonset must follow after the (local) sunset, and not before (i.e. conjunction must take place before local sunset time) on the evening for which the new crescent's visibility is being evaluated.

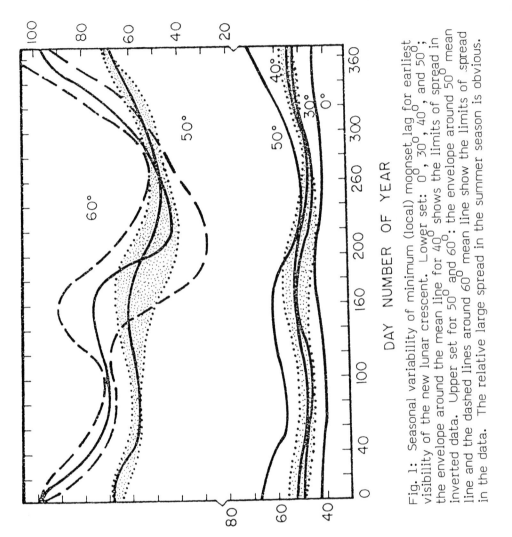

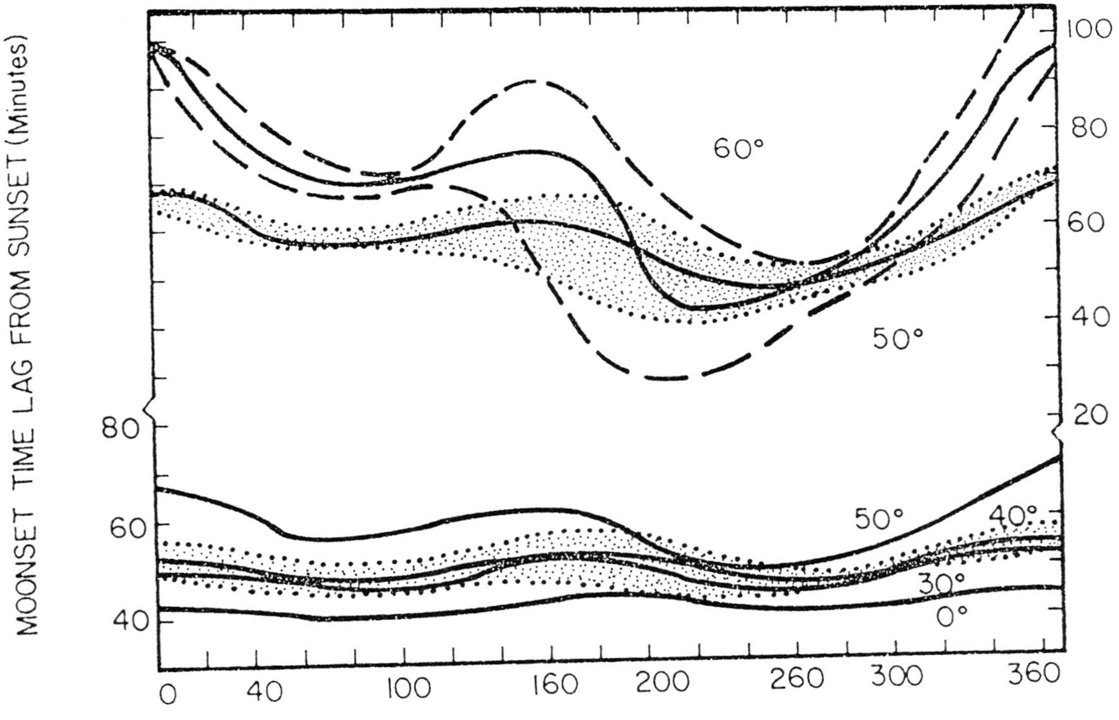

Fig. 1: Seasonal variability of minimum (local) moonset lag for earliest visibility of the new lunar crescent. Lower set: 0^0, 30^0, 40^0, and 50^0; the envelope around the mean line for 40^0 shows the limits of spread in inverted data. Upper set for 50^0 and 60^0: the envelope around 50^0 mean line and the dashed lines around 60^0 mean line show the limits of spread in the data. The relative large spread in the summer season is obvious.

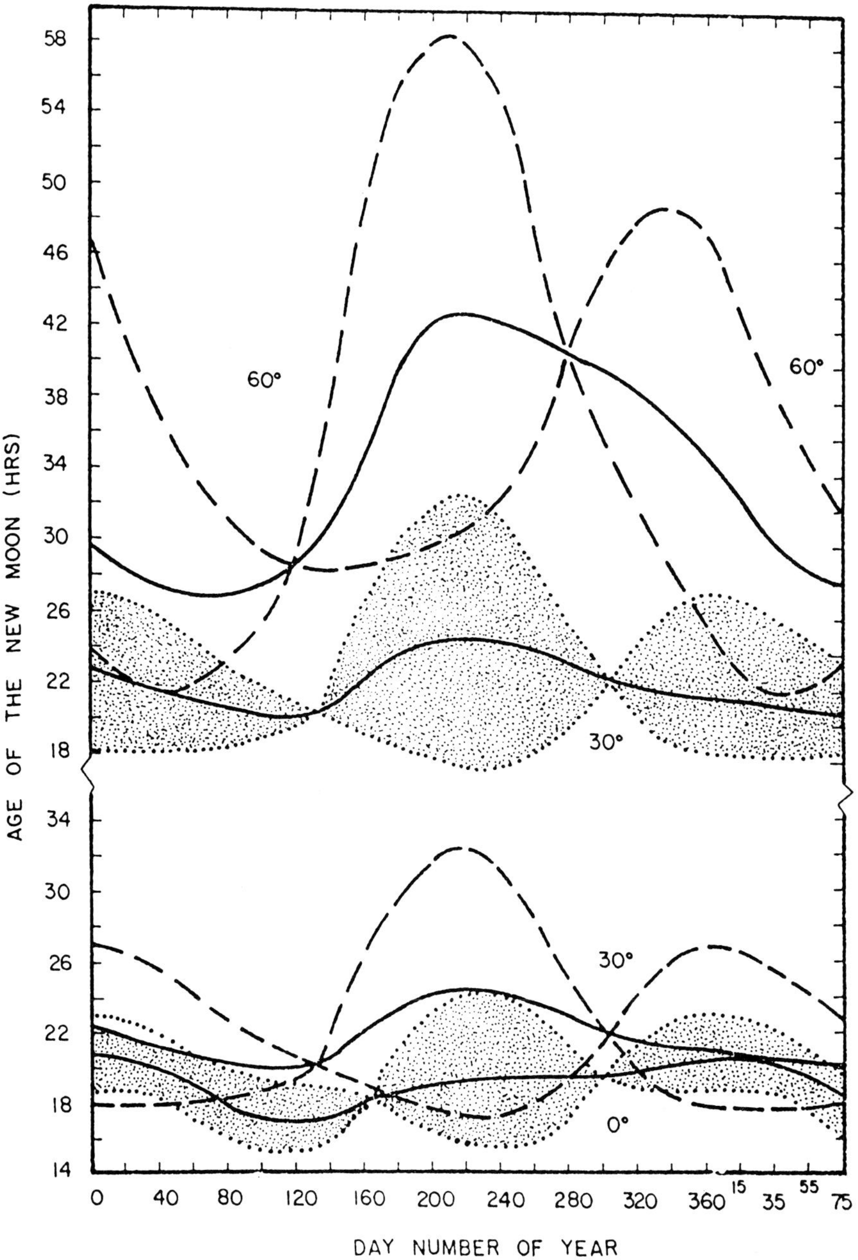

Fig. 2: Seasonal variability of minimum 'moon's age' (at local sunset) requirement for earliest visibility at three latitudes. The envelopes and the dotted lines show the limits of spread in inverted data. The 'age' shows a greater variability and therefore less accuracy of the criterion especially at the higher latitudes.

Appendix: Approximate Diurnal Separation of the Moon

Sun's (apparent) annual motion (360^0 in 365 days)	$= 1^0$ /day
Sun's (apparent) diurnal motion due to earth's rotation	$= 360^0$/day
Sun's net (apparent) diurnal motion	$= 359^0$/day
Moon's (true) monthly motion (360^0 in 29.5 days)	$= 13.2^0$/day
Moon's (apparent) diurnal motion (due to earth's rotation	$= 360^0$/day
Moon's net (apparent) motion	$= (360-13.2^0)$/day $\simeq 14.5^0$/hour
Moon's motion relative to Sun	$\simeq -(13.2-1)^0$/day $\simeq -12.2^0$/day
Hence, the Moon separates from the Sun at an average rate	$\simeq 12.2^0$/day

The Moon moves 14.5^0 per hour on average w.r.t. earth (i.e. a meridian). Hence to cover 12.2^0 -- a separation over one day from the Sun, the Moon would require an extra 0.84 hours (better average: 49 minutes) to cross a meridian compared to the Sun i.e. the Moon would lag behind the Sun by about 50 minutes a day in transiting a meridian and this may be used to determine the time of upper transit, rise or set with respect to the Sun. (We may refer to conjunction or a known instant at which the 'lag' or separation is known).

Reference

Bruin, Franz (1977). Vistas in Astronomy, 21, 331.
Ilyas, M. (1981). Quart. J. Roy. Astron. Soc., 22, 154.
Ilyas, M. (1982) J. Roy. Astron. Soc. Canada, 76, 371.
Ilyas, M. (1983) J. Roy. Astron. Soc. Canada, 77, 214.
Ilyas, M. (1984a). Quart. J. Roy. Astron. Soc., 25, 421.
Ilyas, M. (1984b) A Modern Guide to Astronomical Calculations of Islamic Calendar, Times and Qibla. Kuala Lumpur: Berita Publ., 22 Jalan Liku.
McNally, D. (1983) Irish Astron. Journal, 16, 17.

DISCUSSION

E.S.Kennedy: To what extent does the lunar latitude vitiate a criterion based on the age of the moon ?

M. Ilyas : I suppose, the lunar latitude would be a parameter in actual visibility, hence the lunar age rule is to be regarded as a first approximation only.

4
Mediaeval astronomy

A more detailed analysis of the Oriental material shows that the star nomenclature generally called "Arabic" is composed of two elements : on the one hand names of indigenous Arabic origin, and on the other hand names derived from Arabic translations of ancient Greek sources.

- P.Kunitzsch (p.155)

Shigeru Nakayama and Yukio Ohashi

4.1 AN UNKNOWN ARABIC SOURCE FOR STAR NAMES

Paul Kunitzsch
(University of Munich), Davidstr.17, D-8000 München 81,
West Germany

Arabic star names are well known in two areas: in the Orient itself, i.e. in the Arabic-Islamic civilization, and in the West where many of them were adopted since mediaeval times and continued to be used until today.

The complex known in modern Western astronomy as "Arabic star names" is the result of a historical development of almost exactly one thousand years. In mediaeval times, those names were introduced into Western use by Latin translations of Arabic astronomical and astrological works. Afterwards, since Humanist and Renaissance times, and until this present century, Western astronomers used to pick up more "Arabic" names from philological studies of orientalists who tried to describe and explain the stellar nomenclature of the Arabs and other Oriental peoples. As outstanding examples, I mention the studies of Joseph Scaliger and his follower Hugo Grotius (both printed in 1600) whose nomenclature was borrowed by Johannes Bayer into his star atlas *Uranometria* of 1603; Thomas Hyde's commentary to his edition of Ulugh Bēg's star catalogue (Oxford, 1665) from which Giuseppe Piazzi borrowed a great number of names into the second edition of his Palermo Catalogue, 1814; German studies by F.W.V. Lach (1796) and Ludwig Ideler (1809) which were used by continental astronomers such as J.E.Bode and many others; and still the book on star names by R.H. Allen (1899) from which several new names appear in astronomical books and atlases of our times.

But let us come back to the Orient itself. The star names used throughout in astronomy and astrology in the Arabic-Islamic civilization are generally called "Arabic". A more detailed analysis of the Oriental material shows that the star nomenclature generally called "Arabic" is composed of two elements: on the one hand names of indigenous Arabic origin, and on the other hand names derived from Arabic translations of ancient Greek sources. See for example the name of the star alpha Tauri, Arabic *al-dabarān* (borrowed into the West as Aldebaran), which is a name of old Arabic origin, and on the other hand the name of the star alpha Piscis Austrini, Arabic *fam al-ḥūt al-janūbī* ("the Mouth of the Southern Fish", borrowed into the West as Fomalhaut), which is derived from Ptolemy's description of the position of the star in the star catalogue of his *Almagest*.

In the past three hundred years a great number of Arabic source texts has been studied and edited so that it can be assumed that, today, we have a fairly complete and well founded knowledge of the Arabic star nomenclature, both of its indigenous and the translated Greek-based branches.

As far as the indigenous Arabic star nomenclature is concerned, this has been collected and transmitted down to us by several mediaeval Arabic philologists and lexicographers of whom I mention the following whose related works have already been published: Quṭrub (d. after 825), Ibn Quṭayba (d.884, or 889, in two works), Ibn ʿĀṣim (d.1013), al-Marzūqī (d.1030), Ibn Sīda (d.1066), Ibn al-Ajdābī (d.prior to 1203), and pseudo-Ibn Fāris. To these are to be added the famous scholar al-Bīrūnī (d.1048, in a work written in 1029) and the well-known astronomer Abu'l-Ḥusayn al-Ṣūfī (d.986) who wrote a special work on the fixed stars and the 48 classical constellations according to the *Almagest* tradition in which he also mentioned a great number of indigenous Arabic star names identifying them astronomically with the respective Ptolemaic stars.

As for the 28 lunar mansions, they are part of the indigenous Arabic star traditions (although, ultimately, received from India); they are included in all of the sources mentioned above and are listed and described in many other texts.

For the Greek-based Arabic stellar nomenclature, there is equally a reasonable number of souces available. The point of departure for this branch of nomenclature were of course the Arabic translations from the Greek. The main work to be considered here is Ptolemy's *Almagest* of which several translations were made into the Arabic. Of these, two versions have survived until today, the translation of al-Ḥajjāj ibn Yūsuf ibn Maṭar, and the translation of Isḥāq ibn Ḥunayn with the emendations of Thābit ibn Qurra. A critical edition of these is presently in press. Apart from the *Almagest*, also numerous other Greek texts were translated into Arabic containing individual star names or lists of the 48 classical constellations. Of these I only mention Ptolemy's astrological work *Tetrabiblos* (of which several Arabic translations have survived, in manuscript form, but still unpublished) which in book I, chapter 9 contains a description of the constellations and many prominent stars. Further, the great star catalogues by al-Battānī, al-Ṣūfī, al-Bīrūnī, al-Ṭūsī, and Ulugh Bēg have passed on a good portion of the Greek-based stellar nomenclature. To these must be added the astrological work *Introductorium maius* of Abū Maʿshar, al-Bīrūnī's astrological handbook *Tafhīm*, the *Encyclopaedia of Sciences* by Muḥammad ibn Aḥmad al-Khwārizmī (late 10th century), the *Cosmography* of al-Qazwīnī (d.1283),and the world history (in Syriac) of Barhebraeus (d.1286), all of which contain lists and surveys of the 48 classical constellations.

Apart from these major works, there are innumerable minor or non-specialized Arabic writings which also transmit the Arabic stellar nomenclature, be it in astronomy or astrology, in lexicography or poetry or philology, etc. A case of special interest is the Persian romantic verse epic *Vīs u Rāmīn* by the poet Fakhr al-Dīn Gurgānī (around 1050) which in a kind of a greatly expanded horoscope contains a full description of the 48 Ptolemaic constellations.

In practical usage, no difference was made by astronomers and astrologers between indigenous Arabic and Greek-based Arabic star names: the two types of names mentioned above (*al-dabarān*, alpha Tauri, and *fam al-ḥūt al-janūbī*, alpha Piscis Austrini) were used indiscriminately through all centuries in the Arabic-Islamic world. Equally, in that mixed form they were borrowed into mediaeval Western translations from the Arabic and continued to be used as "Arabic star names" in modern astronomy until today.

Thus it can be affirmed that our present knowledge of the nomenclature of the constellations and the stars used in the Arabic-Islamic world and, subsequently, in the derived Western sources, is rather comprehensive and complete, both under historical and philological aspects.

The mediaeval Western sources using Arabic star names can be divided into two groups: translations made from the Arabic which borrowed the names directly from the Arabic original texts and made them available to Western readers; and Western writings and compilations which gleaned the Arabic names from translated works, or from earlier Western writings or compilations, that means which used these names on the basis of second hand knowledge.

Methodically, therefore, it must be possible to trace all those Arabic names in Western texts back to their ultimate source and their Arabic origin. This rule has proved successful in all my related research work in the past 35 years, with one exception which will be the subject of the following discussion.

In 1246, in Paris, one John of London composed a star table for the astrolabe containing 40 stars with ecliptical coordinates. The table has been edited in 1966 from seven manuscripts (including one manuscript containing a revised version of the table by John of London's disciple Roger Linconus, dated four years later, i.e. 1250). More manuscripts containing the table certainly exist in Western libraries.

While the identity of this John of London is not yet safely established, his authorship of the star table is well attested. The French scholar, Monsieur Fontès, has published in 1897-98 a letter of John of London addressed to a scholarly friend, R. de Guedingue, in which John has answered a number of astronomical and astrological questions of that person and in the course of which he mentions his star observations in Paris, 1246, and the star table which he had composed subsequent to these observations. The same manuscript, some folios later, also contains a copy of the star table. As it seems, John of London was a well-informed and well-read man. In his letter he quotes, among his authorities, Albategni (= al-Battānī), Thesbith (= Thābit ibn Qurra), Ptolemy's *Almagest*, Abrachis (= Hipparchus), Pythagoreus (perhaps an astrological pseudepigraphon), Arzachel (= al-Zarqāllu), and the astrologer Aomar with his work *De nativitatibus* (= ᶜUmar ibn al-Farrukhān).

Two things in John of London's star table are of special historical interest. The first point is of a purely astronomical character. John of London states that his star table is the result of his own star observations which he carried out in Paris by means of an armillary sphere. This statement is corroborated by the fact that the coordinates of his stars, both the longitudes and the latitudes, are not identical with those of the *Almagest*. In mediaeval times it was most usual to construct star tables by merely computing the precessional difference to be added to Ptolemy's longitudes, for the respective epochs, and to retain Ptolemy's values for the latitudes. Such star tables, therefore, can easily be compared directly to Ptolemy's catalogue in the *Almagest*. This is not the case with John of London's coordinates which are throughout more or less different from Ptolemy's. Therefore, his contention to present in his table the results of his own star observations appears to be true and confirmed by the dates.

The second point is that he adds to nearly every single of his forty stars an Arabic name or designation. And here lies our problem, because among those names there are four which cannot be traced in any of all the Arabic original sources known to us until today.

It may be worthwhile mentioning that John of London, through his Table, has introduced into astronomical use a good number of Arabic star names many of which are living on until our present time. The most famous of these names may be Betelgeuse (for alpha Orionis), which in his spelling was *bedalgeuze* (formed, through a misreading, from the Arabic *yad al-jawzā'*, "the Hand of *al-jawzā'*, or Orion"), which in Renaissance times was wrongly explained and transformed into Betelgeuse (with a *t* in the middle instead of the original *d*).

The four names that have remained undocumented in the original Arabic tradition are the following:

Nr.6 in John's table (alpha Arietis), is called in two of the seven manuscripts of the 1966 edition *enif* which is clearly derived from the Arabic word *anf*, "nose". But the Arabic translations of the *Almagest* and all the subsequent Arabic sources use another term, *al-khaṭm* ("the muzzle"), instead, in congruence with Ptolemy's Greek text.

The same word, *enif* (from Arabic *anf*, "nose"), is also used in John's star no.37 (epsilon Pegasi), and in this position the name Enif has survived until our present time. Here again, the Arabic original sources use other terms, *viz. al-jaḥfala* ("the lip", for a horse's lip) in the two versions of the *Almagest*, and *al-ḥulqūm* ("the throat") in al-Battānī's catalogue. These stand in the place of the original Greek term for the "muzzle" of the Horse. In both stars, alpha Arietis and epsilon Pegasi, John has inserted the Arabic word for "nose" contrary to the existing Arabic and Greek traditions which had both stars on the "muzzle" (or the "lip", respectively). No known Arabic text (or inscribed instrument, as an astrolabe or celestial globe) ever used the word *anf*, "nose", in relation to these two stars.

Further, no.9 in John's table, the star gamma Eridani (which was the 10th star in Ptolemy's constellation of Eridanus). Ptolemy described the star as "The rearmost of the four stars in the next interval" (English translation of G.J. Toomer, London, 1984). The word "interval" appears in the Arabic translations as *al-buᶜd* or *al-masāfa* (i.e. "the distance"). But John of London calls the star *algetanar* which seems to be a Latin transliteration of an Arabic form *ᶜarjat al-nahr*, "the Bend of the River". Such a "bend" is not mentioned by Ptolemy for this star, moreover a "bend" occurs with Ptolemy in the descriptions of the 2nd, 18th and 29th stars of Eridanus. Further, the Arabic sources do not use for these "bends" the word *ᶜarja* which was applied by John of London to the 10th star, gamma Eridani. That means both the Arabic word as such and its application to the star gamma Eridani remain undocumented in the sources. By the way, John's mediaeval name has lived on in some modern sources, in the revised spelling Angetenar, as a name for the star tau^2 Eridani (which was the 19th star of the constellation with Ptolemy).

The fourth case is John's star no.17 (rho Puppis, i.e. the 2nd star in Ptolemy's constellation Argo). To this star John adds, in Latin characters, the word *markeb* which doubtlessly is derived from the Arabic word *markab* which, among other things, can designate a "ship". In all the surviving Arabic original sources, however, Ptolemy's ship Argo is unanimously called *al-safīna*, and never anything else. (In modern astronomy the name Markeb is applied to the star kappa Velorum.)

That means that in these four stars John of London has used Arabic terms that never appear in any of the Arabic original sources known to us, some of them even contrary to the Ptolemaic location of the respective stars.

To these four can be added a few doubtful cases. For John's star no.39 (beta Pegasi) two of the seven manuscripts used in the edition of 1966 give a transliteration of its common Arabic, Greek-based name *mankib al-faras*, "the Horse's Shoulder", while the other manuscripts, instead, give a transliteration of a rarer Arabic designation, *yad al-faras*, "the Horse's Forefoot" (spelled *bedalferaz*, with the same misreading as in the name of alpha Orionis, cited above, which was *bedalgeuze*, for Arabic *yad al-jawzā'*, "Orion's Hand"). At least, *yad* for the "forefoot" of the Horse is documented in al-Ḥajjāj's translation of the *Almagest* and in al-Ṣūfī's description of the constellation. But it is surprising to find two entirely different names given in the various manuscripts to the same star.

In star no.7 (alpha Ceti), John applies the Arabic name *menkar* (which is still in use for the same star in our times). This is certainly the original Arabic *al-minkhar*, "the nose", and correctly translated into Latin as *naris ceti* by John of London. The term is well documented in the Arabic astronomical tradition - the problem, however, is that this designation originally was applied to lambda Ceti (the 1st star in Ptolemy's constellation of Cetus), while John gives it to alpha Ceti (which is the 2nd star in the constellation). So, here, the term itself is correct and historically documented, but the location is wrong. Should John have confused the two stars in his source, the *Almagest*, jumping over from the star alpha to the preceding line which contained the description of lamda?

John's star no.31 (gamma Draconis) is called *razcaben*, which is a corruption of *raztaben*. The Ptolemaic constellation of Draco is called *al-tinnīn* ("the Snake") in all the known Arabic sources. John's spelling, therefore, seems to be a corruption of the correct Arabic name for this star, *ra's al-tinnīn*, "the Snake's Head". This seems to be confirmed by the spelling offered in one of the seven manuscripts used for the 1966 edition: *racaten*. Neverthless, in Renaissance times it has been ventured that John's spelling was derived from an assumed Arabic *ra's al-thu^{c}bān*, using, for the constellation, the word *al-thu^{c}bān* (which is undocumented in the original sources) instead of the common and well-documented *al-tinnīn*. I am not inclined to accept the Renaissance interpretation of John's form of the name and prefer its derivation from the singularly and well documented Arabic *al-tinnīn*. But there remains a slight doubt, especially in view of John's four other examples of a totally deviating nomenclature.

Another intricate case is John's star no.1, alpha Cephei, which is called by him *aldramin* (and in one manuscript, more completely, *aldheraymin* - still known today, in a modified Renaissance spelling, as Alderamin); John translated it into Latin as *dextrum adiutorium cephei*, i.e. "Cepheus' Right Arm". All these details do not fit together. In the *Almagest*, alpha Cephei (the 4th star in the constellation) is located in the right "shoulder" (not on the right "arm"). Only the 8th star (iota Cephei) in the *Almagest* is located on the "left arm" where "arm" was translated in the Arabic versions as *al-^{c}aḍud* ("upper arm"), while al-Battānī has *al-sā^{c}id* ("lower arm"), instead. The explanation of John's form *aldramin*, in the sense of "right arm", therefore, remains utterly uncertain. A Renaissance scholar derived John's form from an assumed Arabic designation *al-dhirā^{c} al-yamīn* (in the sense of "right arm"), but this is not documented in the sources as we have just seen, with one exception: Abū Ma^{c}shar, in one instance, mentions Cepheus' left arm, i.e. the star iota Cephei, under the designation *dhirā^{c} qayfawus al-aysar*. But it is rather improbable that John of London saw this remote *topos* in an Arabic copy of Abū Ma^{c}shar's *Introductorium maius*. Further, in the *Almagest*, the word *al-dhirā^{c}* is normally used in the feminine gender so that, in our case, instead of the proposed *al-dhirā^{c} al-yamīn*, the designation should correctly read *al-dhirā^{c} al-yumnā* which would no longer bear an assonance to the Latinized form *aldramin*. Another suggestion,which I have supported all the time and continue to do so, is that the Latinized Arabic term *aldramin* is but a variant out of the many various Latinized spellings of the Arabic name of the star alpha Geminorum, both in astrolabe star lists and in lists of the lunar mansions (where alpha and beta Geminorum form the 7th mansion). This name would have been wrongly transferred to the star alpha Cephei in John's list. The problem cannot be definitely solved until the discovery of further documentary evidence.

To sum up we can state that John of London's star table is of twofold historical interest: astronomically, because the coordinates in the table represent values found by John of London himself, through observation; and philologically, because he has introduced a number of new Arabic star names many of which have lived on in astronomy until our present time.

From the textual descriptions of several stars it is obvious that John has really carried out observations of his own, because he describes these stars quite in his own words, according to his own experience and impression, and different from the traditional descriptions in the *Almagest*.

Further, in many other stars of his list we find literal quotations from Gerard of Cremona's Latin translation of the *Almagest* from the Arabic which proves that Gerard's text was one of John's written sources.

As for the Arabic names added by John to nearly all of the 40 stars in his table, all of them were introduced by him - either names that were already transmitted in earlier Western texts, but for which he found his own new spellings, or entirely new names that were never used before in any Western work derived from Arabic sources and which appear for the first time in Europe in his star table.

It is evident that he could not have gleaned his Arabic names - neither the new spellings of older known names, nor the entirely new names - from a translated Western text. At least, no such text preceding his star table of 1246 has ever become known to us.

The question, therefore, arises whence John obtained his Arabic material, and in which way he utilized it. An additional complication, in this connection, lies in the fact that among his new Arabic material he even offers four names which cannot be traced in all the Arabic original sources, written texts and inscribed instruments (such as astrolabes and celestial globes) known to us until today.

It can hardly be assumed that he himself had a working knowledge of Arabic sufficient to read and evaluate original Arabic text material. Rather it must be assumed that somebody knowing Arabic sufficiently assisted him in the composition of his star table. Through this helper John might have obtained all the Arabic names and terms which he added to his table.

Most of these are identical to the traditional Arabic terminology found in the Arabic translations of the *Almagest*. Therefore, it appears likely that an Arabic copy of the *Almagest*, or an Arabic celestial globe using the *Almagest* nomenclature, was among the sources which John's assistant had at his disposal.

As for those four names that cannot be traced in all the existing Arabic original sources, one possibility is that these were not taken from a written source but rather that John's helper invented them by himself, perhaps in some places where the source was defective, or by mere indolence or carelessness. This possibility appears the more likely when we consider the remarkable stability and uniformity of the nomenclature in all the known Arabic original sources. Further search for the "unknown source" of those four star names of uncertain origin in John of London's star table, therefore, might never arrive at a result because such a "source" might never have existed.

REFERENCES

Fontès,M.(1897-98). ed. (John of London's Letter to R. de Guedingue), Bulletin l'Académie des Sciences Inscriptions et Belles-Lettres de Toulouse, **1**, 146-160, Toulouse.

Kunitzsch,P. (1959). Arabische Sternnamen in Europa, Wiesbaden : Otto Harrassowitz. (On Arabic star names in European modern astronomical use).

- (1961). Untersuchungen zur SternnomenKlatur der Araber, Wiesbaden: Otto Harrassowitz. (On indigenous Arabic stellar nomenclature).

- (1966). Typen von Sternverzeichnissen in astronomischen Handschriften des zehnten bis vierzehnten Jahrhunderts, Wiesbaden : Otto Harrassowitz, pp.39-46 ("Typ VI") (Includes edition of John of London's Star Table).

- (1974). Der Almagest, Die Syntaxis Mathematica des Claudius Ptolemäus in arabisch-lateinischer Überlieferung, Wiesbaden (On the terminology and nomenclature of the Stars and constellations in the Almagest).

- (1977). Arabische Sternnamen - Sternnamen der Araber, Sudhoffs Archiv, **61**, 105-117 (Arabic star names and star names of the Arabs, definitions).

- (1983). Über eine anwā'-Tradition mit bisher unbekannten Sternnamen, Bayerische Akademie der Wissenschaften, Philos.-Hist. Kl. Sitzungsberichte, Heft 5. (On indigenous Arabic Stellar nomenclature).

- (1986). John of London and his unknown Arabic Source, Journal for the History of Astronomy, **17**, 51-57.

Kunitzsch,P. and Smart,T. (1986). Short Guide to Modern Star Names and Their Derivations, Wiesbaden : Otto Harrassowitz.

DISCUSSION

J.A.Eddy : Do you know more about John of London than you allowed yourself time to tell?

P.Kunitzsch : I do not know much more beyond what I have said in the lecture. The editor of J. of L's letter, M.Fontès, assumes that our John of London is identical with one "Jo Lo" mentioned several times in the works of Roger Bacon. But this is not certain. There is a well known astronomer John of London, working 30-40 years later in England, but it can hardly be assumed that our John of London (in Paris 1246) is the same person as that second John of London in England.

O.P. Jaggi, Xu Zhentao and Michel Teboul

4.2 AL'SUPHI'S STAR-ATLASES AND MIDDLE EUROPE

Karl A.F. Fischer
No.84 Rott, F-67-160 Wissembourg,
France.

The oldest European mythological figures of the star constellations originate from the Farnesian Globe of the 3rd century B.C. The "Centaurus" with the Therion "Lupus" is the most important constellation from which I derive all the series specified. Centaurus holds the wolf with one hand on the hind leg; in the other hand he holds a lance with which he cuts the wolf's throat. This constellation is incorrectly drawn in the first publication of the Farnesian Globe by Bianchini. These antique figures were the pattern for Cod. Vind. 5415, probably by John Dorn, a Dominican, who made also the globe of Bylica at the Budinian court of the king Mathias. This globe is at present the property of the University-Museum of Krakau. The antique constellation-forms were the pattern for the celestial maps of Sebastian Sperantius of Nurnberg in 1503. Due to Albrecht Durer these mythological figures of the constellations re-appeared in the renaissance. We can find them by John Honterus, John Middoch, on many globes for example that of Coronelli, or of the manuscript globe of the Bibiliotheca Cassanatense in Rome of Moroncelli from the year 1716. They survived till modern times in the "Uranometria" of John Bayer, in the star-atlases of John Flamsteed, John Erlet Bode as well as in the latest star atlas with allegoric figures of Rudiger-Meisner from the year 1805.

Abd-al Rahman al-Suphi made a careful revision of the star catalogue of Ptolemy. He allocated the stars precise positions and therefore his work is of greater importance than the original Greek star-catalogue. Al'Suphi's catalogue was rewritten several times and from it brass globes were made. His mythological figures came from ancient Greece, but they differ from them in minor details. The "Gemini" have specially intertwined hands, and the "Centaurus" holds with one hand the "Lupus" on a hind leg and a "thyros" like Ptolemy a sacrificial stem with wine tendril, but in European transcriptions creation in the form of a flower in the other one. By the globes and atlases of Arabic origin this flower refers to a creation in the tulip form, by the Bohemian a linden branch, by the French-Italian as well as of the workshop of Wenceslaus IV to a creation of a Bourbon lily. By other Bohemian star-catalogues the flower refers to a fighting mass of hussite soldiers.

The first compilation using material from Al'Suphi's atlas in Latin was made in XIIIth century Sicily at the court of Frederic II of Barbarossa. It is preserved as Codex Arsenal 1036. The star-atlas of the last kings of the Přemyls-dynasty, of Wenceslaus II and the III of Prague was made on the basis of a connection link unknown to us today. It is designated as Cod.Cues 207. The atlas is a pen-and-ink drawing which resembles the bible of Welislaus of Prague.

A precise copy of the star-atlas of Frederic II, is the star-atlas of the emperor Charles IV. of Luxemburgh, who was of French descent. The painter signed it as "Petrus de Guinoldis". The atlas was made for the young Charles of Luxemburg about 1320. He received French education in Italy. Precise copies of this atlas are the following MSS:

1 Cod.Hamilton 557 of the "Kupferstichkabinett" of Berlin Dahlem which is an early work from the drawing workshop of the emperor Wenceslaus IV, also

2 Clm 826, which was a wedding present to Wenceslaus and Eusophia of Bavary and dates from 1392. But it is a collective work art-historically on which one was working till the second decade of the 15th century and it remained unfinished in spite of it. The background of all mythological figures is an inlaid work, the "Auriga" is provided with the initials "W" and "E", that means Wenceslaus and Eusophia.

3 The most recent of this series is dated 1428 and produced by the late art school of Prague, which could not be active in Prague due to the religious reformation war. This atlas is deposited in the State-Library of Gotha. It was published as a coloured edition in the last year. It is a Bohemian work according to the drawing style. The painters of the Prague art school wandered partly to Raudnitz, where they painted a fresco in the Augustinian canon monastery, partly to the powerful catholic South-Bohemian aristocracy of Rosenberg.

In the art-historical estimation I am indebted to my late friend Doc.Josef Krása.

With the example of "Taurus" I would like to demonstrate that the mentioned atlases are identical.

The newest link of this series is the Cod. Vind. 5318, the explicit of which was first written in 1477 in Salzburg. But it is a Bohemian work of the middle of the 15th century, therefore older than the explicit. This atlas with the shaded pen-and-ink drawings is similar to the Cod.B-2 of Zittau, which is also of Bohemian origin.

The star-atlas of the Prague Magister Wenzel Koranda of Pilsen is a link with the other series of atlases. This atlas belonged to the University-Library of Koenigsberg till the end of the second world war, signature Cod. lat. 1735. At present it is deposited in the University Library of Torun as Cod. 74. The star atlas is about 20 years older than the remaining pages of the Codex and originates from 1450. The Centaurus holds a fighting mass of hussite soldiers.

The Al'Suphi's form of mythological pictures disappeared from public view in the 15th century and was republished only once by John Conrad Schaubach: Erastotelmi Catatherismi, Gottingen, published 1795.

These Al'Suphi's forms were not used in all Arabic MSS. The Arabic translation of Aratus for example compiled with the old Aratus of the Cod. Vat. Graec. 1087. The Arabic "Gemini" comply with Al'Suphi's forms, for example MS orient.Qu 704 Berlin, the Sprenger Cod. 1855 Berlin, or the Cod.Arab.Monachensis 870. But the "Centaurus" is affiliated with the so called "Aratus-Series".

The oldest presentation of the "Gemini" of the "Aratus Series" is Cod.Voss.79 of Leiden, which belonged to Charles the Great. The "Gemini" are characterised with a lyra-like musical instrument for example in the Cod.Prag.433, or in the "Wenceslawicum" Cod.Vind.235, or in the atlas of Zittau of the Bohemian Origin Cod.B-2.

The "Centaurus" in the Aratus poems has very different forms:

1 in Cod XIV.-D-37 Napoli. The Centaurus holds two branches only.

2 in the Cod.Cracov.3411 (a Bohemian work). The Centaurus holds a "Therion" a rabbit-like animal on hind legs with both hands.

3 in Cod.Basil.An-IV-18 or Cod.Coloniensis 83. The Centaurus holds a "Therion" on the back.

4 In the Cod Matritensis 19, Cod.Dresdensis 186 and Cod. Prag.1717: The Centaurus has two "Therions" he holds one by hind legs, the other is caught on the lance.

5 The most wide-spread is the form of Centaurus with three "Theorions", the third is a sacrificial beverage. For example: Cod.Montecassiono 3. Cod.lat.mon. 10.268, the "Wenceslawicum" of Vienna Cod.Vind.2352. Cod.Cracov.573 a "Bohemicum" Cod.Vind.2378, Cod.Prag. 433, known as "Liber Sigismundi de Hradec Reginae" and the very beautiful and newestatlas of the Johanitten monastery of Zittau Cod.B-2. of Bohemian order-general Petrus de Teyn.

6 Different is the presentation of the Centaurus of Klosterneuburg, Cod.CCL-625 who holds a spear in the hand.

7 In the Warsowian Cod.Baworowski 498 has the Centaurus four "Therions", the fourth is a bird, which sits on the lance.

I have tried to create a new classification of mythological constellation forms, unlike Boll and Thiele, on the example of the "Gemini" and "Centaurus".

Shigeru Nakayama, Quan Hejun, Michel Teboul
and Xu Zhentao (from left)

4.3 A STUDY OF SOME ASTRONOMICAL DATA IN MUSLIM CALENDAR

Chen Meidong
Institute for the History of Natural Science
Academia Sinica, Beijing,
China

The Muslim Calendar spread into China in 1385 where it was immediately translated into Chinese by the astronomer Yuan Tong and came into use. In 1477, it was further translated by the astronomer Bei Lin and compiled into the "Qi Zheng Tui Bu", a work more or less the same in substance with the Muslim Calendar recorded in the "Ming Shi Li Zhi", both being works of the same source. They left for us the valuable data of the results of research of ancient Arabian astronomers.

On different occasions in the Muslim Calendar, values different with one another are used for the same kind of data. In that case, which of them are used for them are accurate values surveyed and calculated by people who originally worked out the Muslim Calendar? And how are these values calculated from data now available? Now let us take as an example the average daily degree of solar motion (V) relative to the vernal equinox and give a brief illustration:

In "the day table", there are 3 different values for V: The daily degree of motion 59'08" ($V_1 = 0^0.98555556$); $V_2 = 28^0 35'02''/29 = 0^0.98565134$/day; $V_3 = 29^0 34'10''/30 = 0^0.98564815$/day. According to V_1, there are $28^0 34'52''/29$ days and $29^0 34'00''/30$ days.

Comparing with V_1, V_2 and V_3 are lower by 10", which explains why 1" is added to each of the degrees of motion for the 10 dates in the table. This shows that in the minds of those who compiled the calendar, V_1 is a value on the low side, while V_2 and V_3 are better values after revision.

In "the month table", there are also two different values for V: in the case of a common year ($V_4 = 348^0 55'09''/354$ days $= 0^0.98564736$/day); in the case of a leap year ($V_5 = 349^0 54'17''/355$ days $= 0^0.98564710$/day). According to V_2 and V_3, the sun revolves $348^0 55'12''$ in a common year, and $349^0 54'20''$ in a leap year, which are 3" higher than those recorded in the table. It can explain why 1" is made less from each of the degree of motion per 3 months in the table. This shows that in the compiler's opinion, V_2 and V_3 are values on the higher side, while V_4 and V_5 are better values after further revision.

It is known that in 30 lunar years there are 19 common years and 11 leap years with 10631 days in all. According to V_4 and V_5, the total degrees of solar motion in 30 lunar years are $348^0 55'09''$x19+ $349^0 54'17''$x11 = 29x360^0 + $38^0 24'58''$. But in "the 30 lunar years table", calculation is made with 30 lunar years moving $38^0 25'01''$, compared with which our calculated value is lower by 3". It is why 1" is added to each of the degree of motion for the 3 years in the table. This shows that in the minds of those who prepared the calendar, the value more accurate than V_4 and V_5 should be V_6 = (29x360^0 + $38^0 25'01''$)/10631 = 0^0.98564735/day.

If calculated according to V_6, the solar motion of 840 lunar years should be (29x360^0 + $38^0 25'01''$)x28 = 814x360^0 + $355^0 40'28''$. But in "the 1440 lunar years table", calculation is made with 840 lunar years moving $160^0 05'33''$ - $164^0 25'19''$ = $355^0 40'14''$, which is 14" less than in our calculation. It explains why 1" is made less from each of the degree of motion for the 14 years in the table. This shows that in the mind of compilers, V_6 is a value slightly higher, and the real accurate value ought to be V_7 = [(29x360^0 + $38^0 25'01''$) x 28-14"]/28x10631 = 0^0.985647333/day.

In brief, tables such as "the day table", "the month table", "the 30 lunar years[11] and "the 1440 lunar years" in the Muslim calendar are all worked out on the basis of the accurate value V_7, surveyed and calculated by calendar-makers. The various values from V_1 to V_6 are all approximate values of V_7, the only difference being their degree of approximation. This conclusion can be applied without exception to the following astronomical data (1) to (6) whose corresponding value of V_7 can be obtained in a way similar to the one stated above.

1 The average daily degree of solar motion relative to the vernal equinox V_7^t (length of the tropical year):

It is known that V_7^t = 0^0.985647333, $360^0/V_7^t$ = 365^{α}.2421997, this is length of the tropical year. In comparison with the theoretical value for the year 622 and 1348, their respective error is found to be Δ = 6^s.7(2^s.7). The errors of the following astronomical data as against the theoretical values for the year 622 and 1348 are all expressed in this form. However, the error of the Huang Ji Calendar (604) and the Shou Shi Calendar (1281) is found to be Δ = 196^s and 23^s, these are far worse than that of the Muslim Calendar.

Most scholars in the past were inclined to act according to the Muslim Calendar with intercalary 31 days in 128 tropical years, and calculated the length of the tropical year as 365(31/128) = 365^{α}.2421875, Δ = 7^s.8(3^s.7), which is in fact less accurate than the value originally surveyed and calculated by the Arabian astronomers. The intercalary 31 days in 128 tropical years are approximate values of the intercalary 31 days in 127.9936 tropical years.

2 The average daily degree of lunar motion relative to the star V_7^s (length of the sidereal month):

$V_7^s = [(389 \times 360^\circ + 38^\circ 15')\times 28 + 5')/28 \times 10631 = 13^\circ.17639479$, $360^\circ/V_7^s = 27^d.3215858$, this is length of the sidereal month, $\Delta = 6^s.3$ $(6^s.5)$.

However, the error of the Huang Ji Calendar and the Shou Shi Calendar is found to be $\Delta = 1^s.3$ and $1^s.0$, they are more accurate than the Muslim Calendar.

3 The average daily degree of lunar motion relative to the sun V_7^m (length of the synodic month):

$V_7^m = [(719 \times 360^\circ + 359^\circ 40')\times 28 + 8']/28 \times 10361 = 24^\circ.38149482$, $720^\circ/V_7^m = 29^d.5305930$. This is length of the synodic month, $\Delta = 0^s.6(0^s.5)$. However, the error of the Huang Ji Calendar and the Shou Shi Calendar is found to be $\Delta = 0^s.8$ and $0^s.5$, they are of an equal level with Muslim Calendar.

Most scholars in the past were inclined to act according to the Muslim Calendar with the intercalary 11 days in 30 lunar years, and calculated the length of the synodic month as $(30 \times 354 + 11)/30 \times 12 = 29^d.5305556$, $\Delta = 2^s.7(2^s.8)$. In fact, it is approximate values that are used for originator of Muslim Calendar.

4 The average daily degree of lunar motion relative to the apogee V_7^a (length of the anomalistic month):

$V_7^a = (385 \times 360^\circ + 293^\circ 47')\times 28 + 5']/28 \times 10631 = 13^\circ.06497849$, $360^\circ/V_7^a = 27^d.5545804$, this is length of the anomalistic month, $\Delta = 1^s.4(2^s.1)$. However, the error of the Huang Ji Calendar and the Shou Shi Calendar is found to be $\Delta = 0^s.8$ and $3^s.7$, the Muslim Calendar is worse in accuracy than the Huang Ji Calendar and a bit better than the Shou Shi Calendar.

5 The average daily minute of the regression of nodes of ecliptic and lunar orbit V_7^Ω (length of the nodical month):

$V_7^\Omega = (562^\circ 58' \times 28) + 12'/28 \times 10631 = 3'.177351949$, $360^\circ/(V_7^s + V_7^\Omega) = 27^d.2122200$. This is length of the nodical month, $\Delta = 0^s.4(0^s.1)$. However, the error of the Huang Ji Calendar and the Shou Shi Calendar is found to be $\Delta = 0.^s6$ and $0.^s.5$, they are pretty much the same with Muslim Calendar.

6 The average daily degree of motion of the planets relative to the sun (V_7^1-V_7^5) ---the planet synodic period:

Saturn: $V_7^1 = [(28\times360^0 + 42^0 16')\times28+8']/28\times10631 = 0^0.952146686$, $360^0/V_7^1 = 378^d.0930028$, $\Delta = 95^s.7$.

Jupiter: $V_7^2 = [(26\times360^0+234^0 39')\times28+11')/28\times10631 = 0^0.902516842$, $360^0/V_7^2 = 398^d.8845230$, $\Delta = 41^s.1$.

Mars: $V_7^3 = [(13\times360^0+227^0 01')\times28+10')/28\times10631 = 0^0.461576543$, $360^0/V_7^3 = 779^d.9356461$, $\Delta = 38^s.4$.

Venus: $V_7^4 = [(18\times360^0+74^0 15')\times28+12']/28\times10631 = 0.^0616523106$, $360^0/V_7^4 = 583^d.9197207$, $\Delta = 143^s.0$.

Mercury: $V_7^5 = [(91\times360^0+267^0 44')\times28-6']/28\times10631 = 3^0.10673782$, $360^0/V_7^5 = 115^d.8771744$, $\Delta = 26^s.2$.

The error of synodic period of Saturn, Jupiter, Mars, Venus and Mercury in the Da Ye Calendar (607) is found to be: $144^s.7$, $168^s.9$, $909^s.6$, $19^s.3$ and $166^s.9$, and the Shou Shi Calendar $25^s.5$, $349^s.6$, $612^s.7$, $1622^s.3$, and $127^s.7$. In point of the overall level of accuracy of the planet's synodic period, the Muslim Calendar is far better than traditional Chinese calendars. Nevertheless, the latter have their own merits. For instance, in accuracy the Muslim Calendar is not as good as the Da Ye Calendar for Venus and the Shou Shi Calendar for Saturn.

7 The annual value of advance of the Sun's apogee and the aphelion of the planet.

According to Muslim Calendar, the advance of the Sun's apogee and the aphelion of planet in 840 lunar years should be all $12^0 36'55''$, their daily value $V_\alpha = 12^0 36'55''/28\times10631$, $V_\alpha \times V_7^t = 60''.01$, this is annual (tropical year) value of advance of the Sun's apogee and the aphelion of Saturn, Jupiter, Mars, Venus, and Mercury, their error each ought to be 2" and 10", 2", 6", 9" 4". Ancient traditional Chinese Calendar had no conception of the advance of sun's apogee so that it cannot be compared with the Muslim Calendar. As for the annual value of advance of aphelion of the planets, we must wait for appearance of the Da Yan Calendar (728) to give for the first time a quantitative description, but it is far backward in accuracy than the Muslim Calendar. When we come to the Shou Shi Calendar, their error each ought to be: 69", 53", 44", 2", 3". It is rather accurate for Venus and Mercury, but as a whole, it still lags behind the Muslim Calendar.

8 The longitude of the Sun's apogee and the aphelion of Planet. According to Muslim Calendar, at the epoch (16.5 July 622) the longitude of the Sun's apogee and the aphelion of Saturn, Jupiter, Mars, Venus, Mercury should be each:

$W_{Su} = B_1 + 89^\circ 21' = 78^\circ 40' 32''$, $\Delta = 0^\circ.65$

$W_{Sa} = B_1 + 254^\circ 48' = 244^\circ 07' 32''$, $\Delta = 2^\circ.08$

$W_J = B_1 + 180^\circ 08' = 169^\circ 27' 32''$, $\Delta = 2^\circ.88$

$W_{Ma} = B_1 + 135^\circ 04' = 124^\circ 23' 32''$, $\Delta = 6^\circ.33$

$W_V = B_1 + 77^\circ 06' = 66^\circ 25' 32''$, $\Delta = 134^\circ.41$

$W_{Me} = B_1 + 216^\circ 17' = 205^\circ 36' 32''$, $\Delta = 30^\circ.47$

where

$B_1 = 360^\circ - 10^\circ\ 40' 28'' = 349^\circ 19' 32''$

However, at its epoch the error of longitude of the Sun's apogee and the aphelion of planet in the Shou Shi Calendar is found to be each: $0^\circ.60$ and $0^\circ.73$, $2^\circ.80$, $5^\circ.84$, $136^\circ.72$, $1^\circ.16$. With the exception of Venus, the Muslim Calendar is inferior.

9 Mean longitude of Sun and planet. According to Muslim Calendar, at the epoch (16.5 July 622) the mean longitude of Sun and planet should be each:

$\bar{W}_{Su} = 116^\circ 05' 08'' - 01' 04'' = 116^\circ 04' 04''$, $\Delta = 1^\circ.14$

$\bar{W}_{Sa} = \bar{W}_{Su} - 359^\circ 18' = 116^\circ 46' 04''$, $\Delta = 0^\circ.45$

$\bar{W}_J = \bar{W}_{Su} - 145^\circ 19' = 330^\circ 45' 04''$, $\Delta = 0^\circ.90$

$\bar{W}_{Ma} = \bar{W}_{Su} - 264^\circ 06' = 211^\circ 58' 04''$, $\Delta = 1^\circ.13$

$\bar{W}_V = \bar{W}_{Su} + 45^\circ 29' = 161^\circ 33' 04''$, $\Delta = 2^\circ.68$

$\bar{W}_{Me} = \bar{W}_{Su} + 85^\circ 34' = 201^\circ 38' 04''$, $\Delta = 1^\circ.38$

However, at its epoch the error of mean longitude of Sun and Saturn, Jupiter, Mars, Venus, Mercury in the Shou Shi Calendar should be each: $0^\circ.99$ and $0^\circ.36$, $0^\circ.51$, $0^\circ.73$, $0^\circ.02$, $10^\circ.34$. With the exception of Mercury, the Muslim Calendar is inferior.

In summing up what is stated above, we might say that the accuracy of the length of the tropical year set by the Muslim Calendar is much higher than that set by traditional Chinese calendars. What is more is that the conception and value of advance of apogee of the Sun are lacking in the latter. Judging from the overall level of accuracy for determining the synodic period of planet and the value of advance of aphelion, the Muslim Calendar is also better than traditional Chinese calendars. But traditional Chinese calendars are more accurate than the Muslim Calendar in determining the value of the sidereal month. Judging from the overall level of accuracy for determining the mean longitude of Sun and planet and the longitude of Sun's apogee and aphelion of planet, the traditional Chinese Calendar is also better than Muslim Calendar. As for the accuracy in the determination of the value for the synodical month, the anomalistic month and the nodical month, traditional Chinese calendars and the Muslim Calendar are in equal level. All these show that both of them have originality and are all remarkable ancient astronomical works.

REFERENCES

Bei Lin, "Qi Zheng Tui Bu" (七政推步), 2, Commercial Press, Beijing.

Chen Meidong (1983). On the Determination of the Length of the Years and the Months in Ancient China, Collected Works for the History of Science and Technology, Shanghai Scientific and Technical Press, Shanghai.

Chen Meidong (1985). Measurement and Calculation of Longitudes of Planetary Perihelion and Their Values of Advance in Ancient China, Studies in the History of Natural Sciences, No.2, Beijing

History of Sui Dynasty (隋書), 17, 88, Chinese Book Publishing House, Beijing, (1973).

History of Ming Dynasty (明 史), 37, 38, Chinese Book Publishing House, Beijing,(1974).

History of Yuan Dynasty(元 史), 55, Chinese Book Publishing House, Beijing ,(1976).

4.4 ON THE SOLAR MODEL AND THE PRECESSION OF THE EQUINOXES IN THE ALPHONSINE ZĪJ AND ITS ARABIC SOURCES

Julio Samsó
University of Barcelona, Barcelona, Spain

Alphonse X, King of Castille (1252-1284), sponsored astronomical work previous to the compilation of the Alphonsine Tables, carried out by his two Jewish collaborators Yehudah ben Mosheh and Isaac ben Sid. On the one hand, these tables can be considered the first European attempt to develop original research in astronomy, but on the other hand, an analysis of this work should be concerned, first of all, with its Islamic precedents. This is why I tend to consider the Alphonsine Tables as another zīj which should be studied in the light of what we know concerning the development of Islamic zījes in Medieval Spain. Of these, two seem to be well documented as having been known and used by the Alphonsine collaborators. One of them is al-Battānī's zīj, the canons of which were translated into Spanish (Bossong, 1978), and which may have been used to compute the solar positions for the end of each month appearing in four Alphonsine works (Astrolabio redondo, Cuadrante para rectificer, Relogio dell agua, and Lámina Universal). These positions have been computed, however, with the Alphonsine Tables themselves. The second zīj known in the Alphonsine circle was al-Zarqālluh's Toledan Tables, used -- as O. Gingerich has established -- to compute the solar and planetary positions in the horoscope which was cast to establish the propitious moment to start the Latin translation of Ibn Abī-l-Rijāl's Kitāb al-bāriᶜ fī aḥkām al-nujūm (Hilty, 1954, lxii-lxiii).

The two aforementioned zījes might have influenced the Alphonsine work. The Toledan Tables could have been the model for the first draft of the Alphonsine zīj, of which we know only the Spanish canons (Rico IV, 119-183), for in this latter text it is clearly established that the tables were used to compute sidereal longitudes. On the other hand, al-Battānī's zīj might have caused a change of attitude in the Alphonsine collaborators as reflected in the Latin numerical tables, the known canons of which seem to be the work of Parisian astronomers of the fourteenth century, used to compute tropical longitudes (Poulle, 1984).

The prologue to the Spanish canons establishes clearly that observations were made during the period 1263-1272. In spite of this fact it is easy to see that most of the planetary parameters used derive either from Ptolemy or from al-Battānī. The solar parameters seem more interesting however, and this should be related to the insistence of the Spanish canons on solar observations made during more than one year using both the Ptolemaic method (passage of the sun through the equinoxes and solstices) and the method established by the astronomers of the Caliph al-Maᵓmūn (passage of the sun through the mid-points between equinoxes and solstices). These Alphonsine observations may have been the origin, at least in part, of the solar parameters appearing in the *Alphonsine Tables* or in other Alphonsine sources. I will review them briefly

1. *Tropical year*: there is no need to insist on this new parameter (365 days 5;49,15,58,58,56,38,24 hours), for Price (1955,*a* and *b*, pp. 104-107) clearly established its origin.

2 *Solar equation*: 2;10° is the maximum solar equation in the Alphonsine *zīj* and the characteristics of the table imply that it was computed using Ptolemaic methods and not according to the old Indian solutions by sines and by declinations. Attempts to recalculate the table do not, however, give good results: only two values correspond exactly to the recomputed ones, and in fourteen instances the error amounts only to 1". In the rest of the table the error is greater, with a maximum of 30". The Alphonsine astronomers do not seem to have done very good work here.

The Alphonsine maximum solar equation (2;10°) is an original parameter which may have resulted from observation. Jean de Murs, however, seems to consider that the aforementioned parameter is only an adaptation of the value used by al-Zarqālluh in the *Toledan Tables* (1;59,10°) due to the fact that the latter tables give sidereal longitudes, while the longitudes computed with the Latin *Alphonsine Tables* are tropical. (Poulle, 1980, 262-263). I cannot see much sense in Jean de Murs' argument, but I do not feel too sure about the originality of the Alphonsine parameter, for it does not differ too much from others we find in the Hindu-Iranian astronomical tradition: 2;13°, and 2;14° in the *Zīj al-Shāh* (Kennedy, 1958, p. 259) as well as 2;14° in al-Khwārizmī's *zīj* (Neugebauer, 1962 *a*, pp. 95-96); on the other hand we also know that, in the 8th c., al-Fazārī used 2;11,15° and 2;14° and al-Bīrūnī ascribes to the *Sindhind* tradition the use of 2;10,46,40° a parameter that might be related to the 2;10,31° used in the *Paitāmahasiddhānta* (Pingree, 1970, pp. 110-114 and 1968, pp. 103-104). We could, therefore, have here a revival of an old Indian parameter.

3 *Solar apogee, precession, and trepidation*: The *Alphonsine Tables* give us a set of solar and planetary apogees which do not seem to derive from any of the tables in use in Medieval Spain. They should, therefore, be considered original unless the contrary can be proved and this, of course, applies specially to the solar apogee. This set of longitudes has, however, a peculiar characteristic for they are neither sidereal nor tropical. It is easy to prove that they incorporate the constant term of precession based on a revolution of 360° in 49 000 years which is equivalent to a precession of about 26.45" per year. Therefore, if we want to know the actual tropical longitude of the solar apogee for the Alphonsine era (midday of 31.05.1252) we should add to the longitude given in the table of *Radices* (Poulle, 1984, p. 124) the amount corresponding to the trepidational term of precession according to the Alphonsine model:

Radix	80;37,66°
Trepidation	8;03,06°
Solar apogee	88;40,06°

We should remark here that the *editio princeps* (1483) of the *Tables* gives the value of the trepidational term for each one of the different eras used, and that it states that the corresponding value for the Alphonsine era amounts to

$8;04,01^{\circ}$. This small mistake will acquire a certain interest later on.

The trepidation theory used in the Latin version of the Alphonsine Tables has been studied by Delambre (1819), Dreyer (1920), Price (1955, a and b, p. 104-107) Dobrzycki (1965), and Mercier (1967-77) and does not need to be explained here. It is enough to say that it uses exactly the same model we find in the Liber de motu octave spere traditionally ascribed to Thābit ibn Qurra. It does not use, however, the same parameters: the period of revolution of the head of Aries along the small equatorial epicycle is 7 000 years (in the Liber de motu, approximately 4 077 Julian years: see Neugebauer, 1962, b, p. 297) and the maximum value of the equation in the Tabula equationum motus accessus et recessus octave sphere amounts to 9° ($10;45^{\circ}$ in the Liber de motu). No more information can be gathered directly from the Alphonsine Tables, but it can be obtained otherwise.

It is a well known fact since Goldstein (1964) that the approximation $10;45^{\circ}$ sin i does not give good results for recalculating the values of the equation of trepidation in the Liber de motu. The same thing happens with the Tabula equationum of the Alphonsine Tables if we use 9 sin i. A much better approximation method was suggested by Dobrzycki (1965 p. 23) with sin 9° sin i, and results which are almost as good can be obtained if we use another approximate formula, given by Mercier (1976 p. 205):

$$\sin \Delta\lambda = \sin i \; \frac{\operatorname{tg} r}{\sin \varepsilon_0}$$

Mercier's formula has the advantage of allowing us, here, to obtain two new parameters; after a certain number of attempts I have deduced that the best results in the recomputation of the table will be obtained using $23;33^{\circ}$ for the mean obliquity of the ecliptic (ε_0, the same parameter as in the Liber de motu) and $3;34,35^{\circ}$ for the value of r, the radius of the small equatorial epicycle ($4;18,43^{\circ}$ in the Liber de motu). These two parameters will be useful, later on, to consider the problem of the obliquity of the ecliptic according to the Alphonsine trepidation model.

Dreyer (1920, p. 247), Price (1955 b, pp. 104-107), and Mercier (1976-77) have underlined the fact that the Alphonsine precession-trepidation model gives good results for the value of precession in the second half of the thirteenth century. I would like, once again, to insist on this point: we have already seen that the longitude of the solar apogee at the beginning of the Alphonsine era (midday of 31.05.1252) is $88;40,06^{\circ}$. It will be easy to compute the position of the solar apogee for the 16.05.16 A.D., the date at which the value of the equation of trepidation was 0° as Mercier (1977, p. 59) has established (not the 18.05.15 A.D. as Dreyer, 1920, p. 247 pretended). The solar apogee for this latter date will be at $71;32,10^{\circ}$. Therefore, in a period of time of, practically, 1236 years the solar apogee has advanced $17;07,56^{\circ}$, a value which corresponds to the increase of longitude due to precession between the two aforementioned dates; this gives us an excellent mean value of precession which amounts to 49.90" per year.

$17;07,56^{\circ}$ can, on the other hand, be rounded to $17;08^{\circ}$ which is precisely the amount by which the longitudes of stars in Ptolemy's Almagest (VII,5) have been increased in the four Libros de la Ochava Espera as well as in the star catalogue of the Alphonsine Tables. No convincing explanation has been given for the use of this parameter, and I would like to suggest here a new hypothesis: The Alphonsine astronomers confused the first year of the reign of Antoninus (137 A.D.), which is the radix date of Ptolemy's catalogue of stars, with the first year of the reign of Tiberius (which, in fact, started in year 14 A.D.) for the Alphonsine Primera Crónica General de España (ed. Menendez Pidal, 1977 I,111) states that at the beginning of the reign of this latter Emperor took place in year 16 A.D. If this hypothesis could be proved to be true, it would definitely support the Alphonsine character of the precession-trepidation model used in the Alphonsine Tables. Let us also remember that the combination of variable trepidation with constant precession seems to continue a traditional idea in Muslim Spain: in the eleventh century Ṣāᶜid al-Andalusī (Blachère, 1935, p.86) and al-Zarqālluh (Millás, 1943-50, pp. 275-276), as well as al-Biṭrūjī in the twelfth century (Goldstein, 1971 I, 89-91 and II, 173-181), when they trace the history of the precession/trepidation theories, state that Theon of Alexandria (4th c. A.D.) had already combined precession and trepidation.

We should mention, finally, that the Alphonsine astronomers not only succeeded in obtaining a good value of precession for their time, but that their model gave good approximations to a certain number of historical determinations of the longitude of the solar apogee. I will only mention here three positions of the solar apogee which are quoted in Ibn Yūnus' Ḥākimī zīj: for year 450 A.D. this source gives 77;55^{0} (Kennedy-van der Waerden, 1963 p. 325) and we obtain 78;08,32^{0} with the Alphonsine Tables (difference -0;13,32^{0}); for year 632 (beginning of the Yazdijird era) Ibn Yūnus gives 80;44,19^{0}, while we can compute 80;46,28^{0} with the Alphonsine trepidation model (the difference amounts only to -0;02,19^{0}) (Caussin, 1804, p. 134). The third instance, however interesting, is less impressive: it corresponds to Ibn Yūnus determination of the longitude of the apogee in 1004 (Caussin, 1804, p. 216) which amounts to 86;10^{0} while we obtain 85;45,35^{0} with the Alphonsine model (difference ∸ 0;24,25^{0}). All this leads me to the suspicion that the Alphonsine astronomers had a copy of Ibn Yūnus' zīj and that they used its historical data to adjust the results obtained with their precession/trepidation model.

4 Obliquity of the ecliptic: One year, at least, of solar observations should, in principle, imply a determination of the value of the obliquity of the ecliptic. In a most striking way, however, the three Latin editions of the Alphonsine Tables which I have been able to examine (1483, 1524, 1553) do not contain a table of the solar declinations, whilst the table of right ascensions seems to have been copied from the Toledan Tables (Toomer, 1968,pp.34-35) or from al-Battānī's zīj (Nallino 1907, II, 61-64) which used an obliquity of 23;35^{0}. On the other hand, the Castillian canons (Rico IV, 153) mention the parameter 23;33^{0}, which is the one used by al-Zarqālluh, and allude to a table of solar declinations (Rico IV, 136 and 179-180).

All this seems most discouraging until one realizes that, in another Alphonsine treatise, the Cuadrante para rectificar its author, Isaac ben Sid mentions an unreported value of the obliquity (23;32,29^{0}) together with two values of the solar declination which are perfectly compatible with the aforementioned parameter. In fact, there is a complete table of solar declinations which has been preserved in another Alphonsine work whose author is also Isaac ben Sid, the

Libro del relogio de la piedra de la sombra (Rico IV, 6). Here the obliquity is 23;32,30^0 (an obvious rounding of 23;32,29^0) add the values of the solar declinations, apart from three obvious mistakes, are, in general well computed. On the other hand the text states that the table is based on observations made "in our time". Therefore it seems that we have here an original parameter which is extremely precise for Newcomb's formula gives us an obliquity of 23;32,11^0 for 1262 (approximately the beginning of Alphonsine observations) and 23;32,04^0 for 1277 (the date in which the Cuadrante para rectificar was written). We may face here the first determination of the obliquity of the ecliptic made in Christian Europe: A little later, in 1290, Guillaume de Saint Cloud obtained a much worse value of 23;34^0 (Poulle, 1980, 261-262). On the other hand we may have here a possible explanation for the origin of the parameter 23;32^0 which Copernicus in the De Revolutionibus ascribes to Prophatius Iudaeus although it is not used in his Almanach (Swerdlow-Neugebauer, 1984, 133).

The above-mentioned parameter for the obliquity of the ecliptic (23;32,30^0) seems to be known only through one other source: Prof. E. S. Kennedy tells me that it appears in an Arabic zīj preserved in Arabic manuscript 6040 of the Bibliothèque Nationale de Paris. The author's name is Abū Muḥammad ᶜAṭāᵓ b. Aḥmad... Ġāzī al-Samarqandī al-Sanjufīnī, and the date of the manuscript is 17.12.1366. It was written in Ho-chou (modern Han-Chia), in the modern Kansu province, on the upper reaches of the Yellow River. I cannot give now any convincing reason for such coincidence although the hypothesis brought forward by my master Professor J. Vernet, (1984) on possible contacts between King Alphonse and the astronomers of the Marāgha school should be considered here for further study. In any case, I would not like to finish this paper without paying some attention to the fact that the Alphonsine parameter for the obliquity of the ecliptic cannot be the result of computation with the three well-known trepidation models in use in Spain: I am referring to the trepidation models in the Liber de motu, in al-Zarqālluh's Treatise on the movement of the Fixed Stars (Millás, 1943-50, pp. 239-243; Goldstein, 1964; Samsó, 1985, u) and in the Alphonsine Tables themselves.

It is easy to disregard entirely al-Zarqālluh's trepidation model for, as Goldstein clearly explained, it gives values of the obliquity of the ecliptic comprised between 23;33^0 and 23;53^0. The Alphonsine parameter is, therefore, below the minimum zarqāllian value. The values of the obliquity of the ecliptic ε implied in the Liber de motu are not so easy to establish, for the preserved Latin text of the work does not explain any straightforward method to compute the value of ε : Mercier (1976, p. 212 and 1977, p.39) has provided us with two exact formulae to calculate the obliquity according to two possible interpretations of the model of the Liber de motu: in one of them (Goldstein, 1964, and North, 1967) the point of intersection of the movable and fixed ecliptics is kept at a distance of 90^0 from the centre of the small equatorial epicycle; according to the second one (Mercier, 1976, p. 210) the intersection of the two ecliptics is kept at a distance of 90^0 from the moving head of Aries. Using Mercier's formulae it is easy to establish that Goldstein-North's constraint gives good approximations for the values we should expect for the time of Ptolemy (23;51,20^0) as well as for the time of the Caliph al-Maᵓmūn (23;33,52^0 according to Bīrūnī, 1962, p. 91). On the other hand, if we use Mercier's constraint, the results will only be acceptable for the time of al-Maᵓmūn. This

I think, is a good reason to prefer Goldstein and North's hypothesis according to which the minimum value for the obliquity of the ecliptic is $23;33^{0}$ which is attained towards 887 A.D. After this date the value of increases, reaching about $23;40^{0}$ in year 1252 A.D., which corresponds to the beginning of the Alphonsine era. We can, therefore, conclude that the Liber de motu was not used to compute the Alphonsine parameter for the obliquity of the ecliptic.

A third possibility is, as we have seen, to use the trepidation model and parameters embedded in the Alphonsine Tables. I prefer here again to use the Goldstein-North constraint and the corresponding Mercier formula, although we would reach similar conclusions with the other hypothesis. The results of my calculations have been plotted against time in Fig. 1 for a period between 16 A.D: (in which, as we have seen $\bar{i} = 0^{0}$) until the middle of the 14th century. It is easy to see not only that the model was not used to compute the Alphonsine parameter but also that the curve only gives us a fairly good approximation to the historical determination of the value of the obliquity made in the time of al-Maᵓmūn. The values established by Ptolemy, al-Zarqālluh ($23;33^{0}$) and the Alphonsine astronomers themselves seem not to have been taken into consideration. This is a serious drawback if we compare it with the results one can get with the Liber de Motu and with al-Zarqālluh's trepidation model. In fact I am inclined to suspect that the coincidence with al-Maᵓmūn's value could be purely casual, and that King Alphonse's astronomers designed their trepidation model having only in mind some historical determinations of the positions of the solar apogee as well as the value of precession established in their time. They seem to have regarded entirely the variation of the obliquity of the ecliptic.

5 A few concluding remarks: I have tried to underline in these pages a few instances in which the Alphonsine Tables seem to depend on an Arabic astronomical tradition which may go back to the origins of Islamic astronomy and draw on Eastern sources giving information on observation techniques established in the time of the Caliph al-Maᵓmūn. It is also possible that the king's collaborators had a copy of Ibn Yūnus' Ḥākimī zīj, and I do not think there is much need to prove the influence of the Liber de motu, the Eastern or Western origin of which is being discussed nowadays. On the other hand, one can appreciate a certain maturity in the Alphonsine treatment of the subjects: new parameters are established and some of them may be the result of new observations. Sometimes one can appreciate a certain originality in the Alphonsine models: such is the case with the combination of precession and trepidation, an idea which, as we have seen, has clear Andalusian precedents, but which never, to my knowledge, produced a set of tables before the time of King Alphonse. In this sense it is most interesting to remark that Naṣīr al-Dīn al-Ṭūsī seems to refer to such a combination in a not very clear passage of his al-Tadhkira fī ᶜilm al-hayᵓa (Ragep 1982, pp. 216-218 and 31-32 of the translation). Here again we have a possibility of a link between Alphonsine astronomy and the Marāgha school.

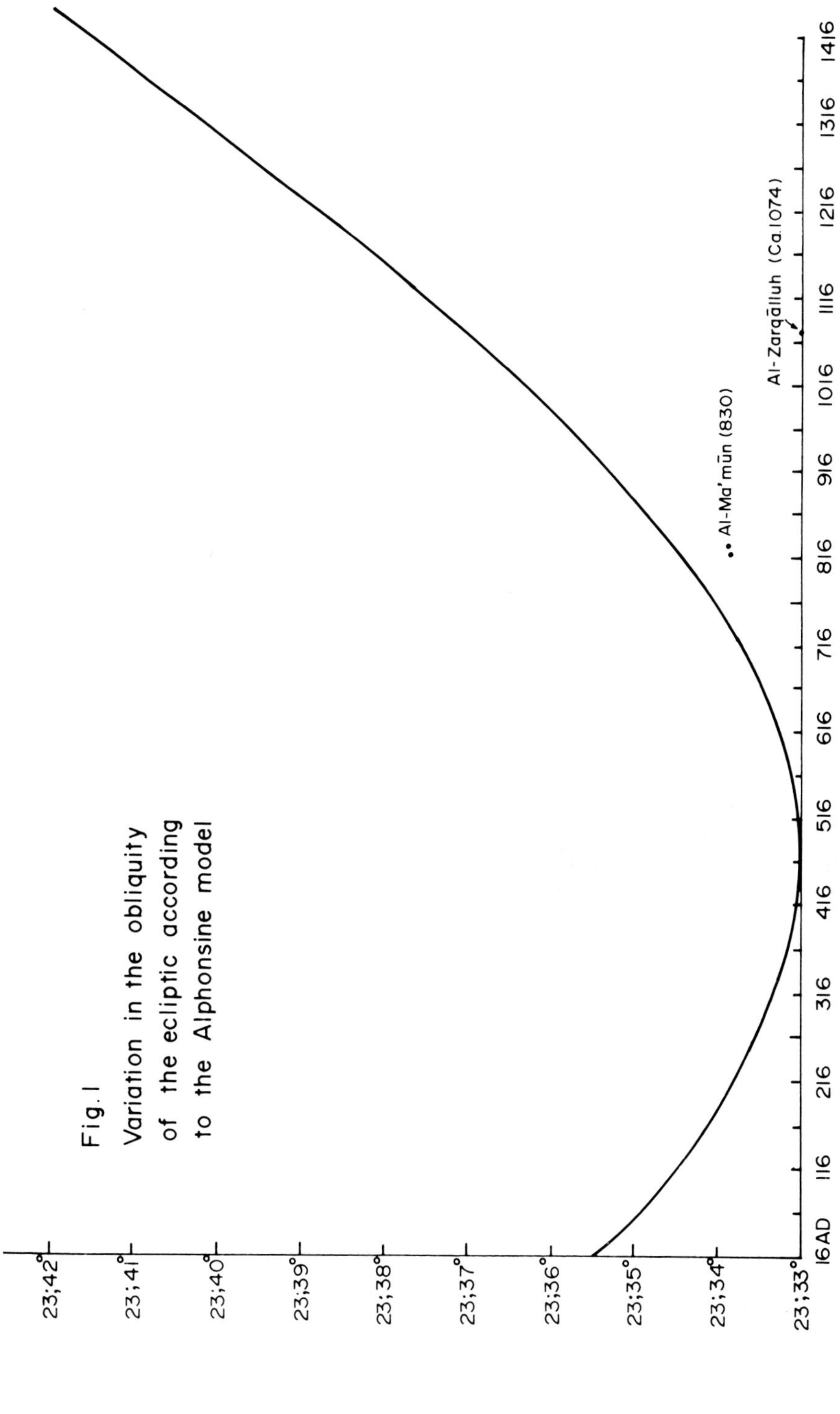

Fig. 1
Variation in the obliquity of the ecliptic according to the Alphonsine model

References

al-Bīrūnī (1962). Taḥdīd nihāyāt al-amākin li-taṣḥīh masāfāt al-masākin, ed. P. Boulgakov. Revue de l'Institut des Manuscrits Arabes, 8.

Blachère, R. (1935). Ṣā^cid al-Andalusī, Kitāb Ṭabaqāt al-Umam (Livre des Catégories des Nations). Pairs: Larose.

Bossong, Georg (1978). Los Canones de Albateni. Tübingen: Max Niemeyer.

Caussin de Perceval, A. P. (1804). Le livre de la grande table Hakémite. Notices et extraits des manuscrits de la Bibliothèque Nationale, [Paris], 7, 16-240.

Delambre, J.B. (1819). Histoire de l'astronomie du Moyen Age. Paris: Courcier. Repr. Johnson 1965.

Dobrzycki, Jerzy (1965). Teoria orecesji w astronomii sredniowiecznej. Studia i Materialy Dziejow Nauki Polskiej, Seria Z,Z. 11, 3-47. (summary in English)

Dreyer, J.L.E. (1920) . The original form of the Alfonsine Tables. Monthly Notices of the Royal Astronomical Society, 80, 243-262.

Goldstein, Bernard R. (1964). On the Theory of Trepidation according to Thābit b. Qurra and al-Zarqāllu and its Implications for Homocentric Planetary Theory. Centaurus, 10, 232-247.

Goldstein, Bernard R. (1971). Al-Biṭrūjī: On the Principles of Astronomy. An Edition of the Arabic and Hebrew versions with translation, analysis and an Arabic-Hebrew-English Glossary. 2 vols. New Haven: Yale University.

Hilty, Gerold (ed.) (1954). Aly Aben Ragel, El Libro Conplido en los Iudizios de las Estrellas. Madrid: Real Academia Española.

Kennedy, E. S. (1958). The Sasanian Astronomical Handbook Zīj-i Shāh and The Astrological Doctrine of "Transit" (Mamarr). Journal of the American Oriental Society, 78, 246-262. Repr. in E.S.Kennedy et al., Studies in the Islamic Exact Sxiences. Beirut: American University of Beirut. 319-335.

Kennedy, E. S. ; B. L. van der Waerden (1963). The World-Year of the Persians. Journal of the American Oriental Society, 83, 315-327. Repr. in (1983) Kennedy et al., Studies in Islamic Exact Sciences, pp.338-350.

Menendez Pidal, Ramón (ed.) (1977). Primera Crónica General de España. 2 vols. Madrid.

Mercier, Raymond (1976-77). Studies in the Medieval Conception of Precession. Archives Internationales d'Histoire des Sciences, 26, 197-220; 27, 33-71.

Millás Vallicrosa, J. M. (1943-50) Estudios sobre Azarquiel. Madrid-Granada: Escuelas de Estudios Árabes.

Nallino, C. A. (1907). Al-Battani sive Albatenii, Opus Astronomicum. Vol. III, Mediolani Insubrum, 1907. Repr. Hildesheim: Olms.

Neugebauer, O. (1962a). The Astronomical Tables of al-Khwārizmi. Translation with Commentaries of the Latin Version edited by H. Suter supplemented by Corpus Christi College MS 283. Hist.-filosofiske Skrifter udgivet af Det Kongelige Danske Videnskabernes Selskab. 4, no. 2, Köbenhavn: Munksgaard.

Neugebauer, O. (1962b). Thābit ben Qurra "On the Solar Year" and "On the Motion of the Eighth Sphere". Proceedings of the American Philosophical Society, 106, 264-299.

North, John D. (1967). Medieval Star Catalogues and the Movement of the Eighth Sphere. Archives Internationales d'Histoire des Sciences, 17, 73-83.
Pingree, David (1968). The Fragments of the Works of Yaᶜqūb ibn Ṭāriq. J. of Near Eastern Studies, 29, 103-123.
Poulle, E. (1980). Jean de Murs et las Tables Alphonsines. Archives d'Histoire Doctrinale et Littéraire du Moyen Age, 241-271.
Poulle, E. (1984). Les Tables Alphonsine avec les canons de Jean de Saxe. Edit. trad. et commentaire. Paris: C.N.R.S.
Price, Derek J. (1955a). A Medieval Footnote to Ptolemaic Precession. Vistas in Astronomy, 1, 66.
Price, Derek J. (1955b). The Equatorie of the Planetis. Cambridge: University.
Ragep, Faiz Jamil (1982). Cosmography in the "Tadhkira" of Naṣīr al-Dīn al-Ṭūsī. Unpublished Ph.D. dissertation. Harvard University.
Rico y Sinobas, Manuel (1866). Libros del Saber de Astronomía del rey D. Alfonso X de Castilla. Vol. IV, Madrid: Impres. ... Real Casa.
Samsó, Julio (1985u). Sobre el modelo de Azarquiel para determinar al oblicuidad de la eclíptica. To be published in the Festschrift which will be presented to Prof. D. Cabanelas (Granada).
Swerdlow, Noel; O. Neugebauer (1984). Mathematical Astronomy in Copernicus' De revolutionibus. 2 vols. New York: Springer.
Toomer, G. J. (1968). A Survey of the Toledan Tables. Osiris, 15, 5-174.
Vernet, Juan (1984). Alfonso X el Sabio: mecánica y astronomía. Conmemoración del Centenario de Alfonso X el Sabio. Madrid: Real Academia de Ciencias Exactas, Físicas y Naturales, 23-32.

Jaipur Observatory. The observatory, popularly known as the Jantar Mantar, was completed in 1734 and has the largest number of Jai Singh's instruments.

4.5 NAṢĪR AD-DĪN ON DETERMINATION OF THE DECLINATION FUNCTION

Javad Hamadani-Zadeh
Sharif University of Technology, P.O.B. 3406, Tehran, Iran

Introduction

A commentary on the spherical astronomy in the Zīj-i Īlkhānī [1], written by the famous Naṣīr ad-Dīn aṭ-Ṭūsī (c. 1270) would be an important contribution to the history of medieval astronomy. But we postpone this and attempt here to describe the contents of the third and fourth sections of Treatise III of this zīj. We hope that it will serve to elucidate a small part of this work, which has remained unexploited. I have made a preliminary English translation of the entire Zīj-i Īlkhānī, which awaits final revision and preparation for publication.

The rules explained verbally and without proof by Naṣīr ad-Dīn in the sections under consideration are similar to the ones described by Kāshī in his Zīj-i Khāqānī and analysed by E. S. Kennedy in [2]. Therefore, we do not present our proofs of the rules, but refer the reader to [2].

The microfilm copy of Manuscript 1418/2 of the Central Library of Tehran University has been used as the main text. It seems to be a careful job, as compared with the other copies available to us: MS 684 of Sipah Salar (now Muṭahari) Library, Tehran, and MSS 5331 and 5332 of the Shrine (Āstān Quds) Library, Mashhad.

Section 2 below presents a description of the contents of the third section of Treatise III in modern notation. It contains the definitions of the declination functions for points on the ecliptic and some trigonometric methods for their determination, Section 3 of this paper describes the material in the fourth section of the text which defines the declination function for arbitrary points on the celestial sphere and gives two formulas for its determination based on spherical trigonometry. We note that the term declination (mayl) was restricted to points on the ecliptic by the medieval astronomers, and they used the word distance (buᶜd) for arbitrary points [2].

The symbols used in the present paper are the standard astronomical symbols, where applicable, and defined as they are encountered. There are no figures in the original Persian text, but we have made them in order to understand the material. To indicate the folios and the corresponding lines of the microfilm copy, we have used numbers inside parentheses. For example, (194r: 2-20) means folio 194r, lines 2-20 of the manuscript.

The medieval trigonometric functions were not identical with their modern counterparts, and are written here with initial capitals to distinguish them from the latter. The relation between the two is, e.g., $\mathrm{Sin}\,\theta = R \sin\theta$, where R is the radius of the defining circle, usually 60, not unity. This explains the appearance of the R in equations below.

Declinations of the Ecliptic Points (194r: 2-20)

The obliquity of the ecliptic (ϵ, ghayat-i mayl) is defined to be an arc of the great circle that passes through the four poles. These are the north and south poles of the ecliptic and those of the celestial equator. The arc between the summer solstice point M and the equator is northern, and the one between the winter solstice point Y and the equator is southern. (See Figure 1 below).

Figure 1

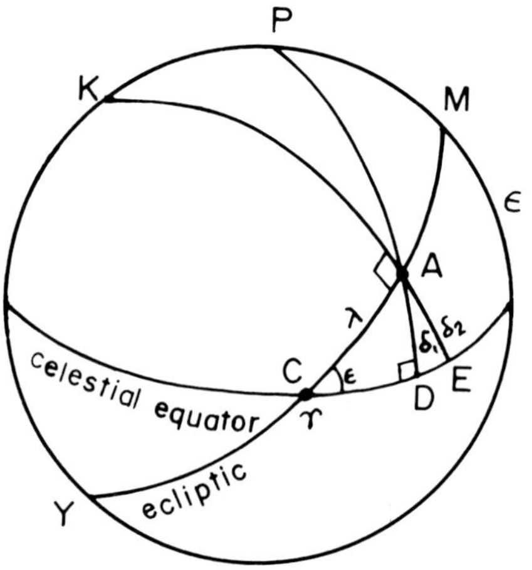

The obliquity of the ecliptic ϵ is determined by observation (line 4). Also, according to our text, Ptolemy's value is $23;51^{\circ}$; most Islamic astronomers give $\epsilon = 23;35^{\circ}$, and some of their successors have given $\epsilon = 23;33^{\circ}$. Naṣīr ad-Dīn prefers his own value, given as $\epsilon = 23;30^{\circ}$, which he adopts.

Partial declinations are defined for an arbitrary point A on the ecliptic, other than the solstitial points (lines 6-10). The first declination (δ_1, mayl-i awwal) is defined as the great circle arc issuing from A and perpendicular to the celestial equator (Fig. 1). This circle is called the declination circle and it passes through the north pole P.

The second declination (δ_2, mayl-i thānī) of the ecliptic point A is the great circle arc indicated as such on Figure 1, issuing from A, as before, but now passing through the poles of the ecliptic. This circle is called the latitude circle, since the celestial latitude (ᶜarḍ) of a star on the celestial sphere is defined as the distance between it and the ecliptic along such a circle.

According to our text (lines 10-11), some medieval astronomers called the first declination the absolute declination, and the second declination the latitude of the ecliptic point A.

In our notation, the text says (lines 11-13) that

$$\delta(\lambda) = \delta(180^0 - \lambda) = -\delta(\lambda + 180^0) = -\delta(-\lambda).$$

As examples, the beginnings of Taurus, Pisces, Virgo, and Scorpio have the same declinations except for a plus or minus sign. The declinations of Taurus and Virgo are northern; those of the other two are southern.

We note that the above equations suffice to determine only the declinations of the points of the first quadrant of the ecliptic which is pointed out in line 13 of our text.

With known λ and ϵ, Naṣīr ad-Dīn uses the right spherical triangle ADC to obtain δ_1. The formula he gives (lines 14-15), in our notation, is the following:

$$\operatorname{Sin}\lambda \cdot \frac{\operatorname{Sin}\epsilon}{R} = \operatorname{Sin}\delta_1, \qquad (1)$$

or

$$\delta_1 = \operatorname{Sin}^{-1}[\operatorname{Sin}\lambda \cdot (\operatorname{Sin}\epsilon)/R]$$

which can be read from the sine table of the zīj.

Given the same values, Naṣīr ad-Dīn uses the right spherical triangle EAC and the following formula (line 16)

$$\operatorname{Sin}\lambda \cdot \frac{\operatorname{Tan}\epsilon}{R} = \operatorname{Tan}\delta_2 \qquad (2)$$

to obtain δ_2 as

$$\delta_2 = \operatorname{Tan}^{-1}[\operatorname{Sin}\lambda \cdot (\operatorname{Tan}\epsilon)/R].$$

Then, using the same triangle, he first obtains the angle $\widehat{AEC}$ (line 17-18) from

$$\operatorname{Cos}\lambda \cdot \frac{\operatorname{Sin}\epsilon}{R} = \operatorname{Cos}\widehat{AEC}$$

and, having calculated Cos $\widehat{AEC}$, he uses it in the following formula to obtain Cos δ_2 as

$$\operatorname{Cos}\delta_2 = \frac{\operatorname{Cos}\widehat{AEC}}{(\operatorname{Cos}\epsilon)/R}.$$

Then

$$\delta_2 = \operatorname{Cos}^{-1}[(R \operatorname{Cos}\widehat{AEC})/\operatorname{Cos}\epsilon].$$

The last two lines of the third section of our text state that if a table is available for the right ascensions of an ecliptic point λ, which we denote by $A_0(\lambda)$, to be defined later in the text, then λ is found for each value of $A_0(\lambda)$ from this table. Then δ_1 and δ_2 are determined by the above rules. To facilitate the determination of the declinations of the ecliptic arc and the corresponding inverse operations, Naṣīr ad-Dīn has tabulated the tables of both declinations δ_1 and δ_2 for arcs from 0^0 to 90^0 in his zīj.

Declination of Other Points (194r:21-27)

The distance (buᶜd) or the present day declination δ of a point X on the celestial sphere from the celestial equator is defined to be an arc of the declination circle, issuing from X and perpendicular to the celestial equator (Figure 2).

Figure 2

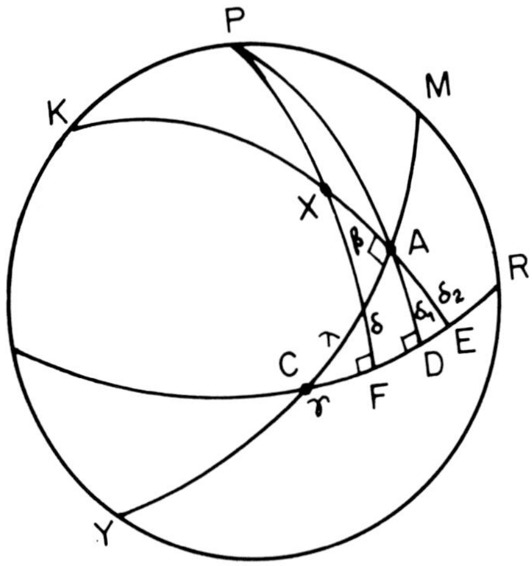

To determine this distance δ = XF, Naṣīr ad-Dīn says (lines 22-24) find the algebraic sum of β , the celestial latitude, and δ_2, the second declination of its celestial longitude λ . This sum, $\beta + \delta_2$, is called the argument (ḥiṣṣah) of the declination, its direction being that of the algebraic sum. Then he determines the first declination δ_1 corresponding to λ, and calls it the inverse declination of the point X. Finally, to determine δ , he prescribes the following formula (line 25);

$$\mathrm{Sin}(\beta+\delta_2)\cdot\frac{\mathrm{Cos}\,\delta_1}{R} = \mathrm{Sin}\,\delta ,$$

from which δ can be determined using the inverse sine function. Alternatively, (lines 26-27) in our symbols, he suggests the following formula:

$$\frac{\mathrm{Sin}\,(\beta+\delta_2)\,\mathrm{Cos}\,\varepsilon}{\mathrm{Cos}\,\delta_2} = \mathrm{Sin}\,\delta ,$$

from which again δ can be determined.

References

[1] Naṣīr ad-Dīn aṭ-Ṭūsī, Zīj-i Īlkhānī. Microfilm 1418/2, Tehran University Central Library. (Other copies consulted are: MS684, Mutahari Library, Tehran, and MSS 5331 and 5332 of the Shrine Library, Mashhad.)

[2] Kennedy, E. S. (1985) Spherical Astronomy in Kāshī's Khāqānī Zīj. Zeitschrift für Geschichte der Arabisch-Islamischen Wissenschaften, 2, 1-46.

DISCUSSION

A.K.Bag : It is referred that Zīj-i-Khāgāni of al-Kāshi is a work on spherical trigonometry. In Indian tradition kha means heaven or celestial sphere, gāni is similar to jyāni which means arcs (may be sine or cosine arcs of) celestial sphere. The rule of three and four were known to India from beginning of early times. This suggests that Khāgāni that is somewhat similar to Khajyāni which means arcs of the celestial sphere. Since both sine and cosine functions were known in India from Āryabhaṭa's time, was it that the al-kāshi was influenced by knowledges from India.

Javed Hamadani Zadeh : I do not think that the name khāgāni in any way related to the traditional words in Sanskrit or Indian language.

The sine function was perhaps taken from the Indian sources, but the rule of four and the sine law are the works of the Islamic period.

S.M.R.Ansari : Do you think that al-Ṭusi's trigonometry as given in his Zīj is the further development of the subject as given in his other works ?

May I draw your attention to a number of manuscripts of Zīj-i-Ilkhanī with a commentary at Mulla Firoze collection at Bombay ?

Javed Hamadani-Zadeh : No, there was no further development. The spherical trigonometry formulas were known by the readers of the text in his time.

Thank you for your information and will try to locate these manuscripts.

The Great Samrat Yantra. A 23 m high equinoctial sun dial designed to measure local time, and the declination and right ascension of celestial objects.

4.6 A NOTE ON SOME SANSKRIT MANUSCRIPTS ON ASTRONOMICAL INSTRUMENTS

Yukio Ohashi (Visiting Scholar)
Dept. of Mathematics, University of Lucknow, Lucknow, India.
(Permanent address: 3-5-26 Hiroo, Shibuya-ku, Tokyo, Japan)

INTRODUCTION

The earliest astronomical instruments in India are the *śaṅku* (gnomon) and the *ghaṭikā* (clepsydra). The former is mentioned in the Śulbasūtras, and the latter in the *Vedāṅgajyotiṣa*. *Āryabhaṭa* described a rotating model of the celestial sphere. After *Āryabhaṭa*, several instruments were described by Varāhamihira, Brahmagupta, Lalla, Śrīpati, and Bhāskara II. After Bhāskara II, some Sanskrit texts specialized on astronomical instruments were composed. The earliest text of this kind is the *Yantra-rāja* (AD 1370) written by Mahendra Sūri. It is also the first text on the astrolabe in Sanskrit. After Mahendra Sūri, Padmanābha, Cakradhara, Gaṇeśa-Daivajña etc. composed Sanskrit texts on instruments, but most of them remain unpublished.

YANTRA-KIRAṆĀVALĪ OF PADMANĀBHA

Padmanābha composed the *Yantra-kiraṇāvalī* or *Yantra-ratnāvalī* (ca. AD 1400), of which Chapter II entitled *Dhruvabhramaṇa-adhikāra* is well known[1].

The *dhruvabhramaṇa-yantra* is a rectangular board with a slit to observe the "polar fish" (a group of stars around the North Pole) for finding time.

The Tagore Library of Lucknow University has a unique manuscript of its Chapter I, namely the *Yantrarāja-adhikāra*.[2] It consists of 116 verses and has a commentary, probably written by its author Padmanābha himself[3]. It describes the construction and use of an astrolabe. Padmanābha takes the circumference of the instrument as the diurnal circle of the first point of Cancer, and draws the diurnal circles of the first points of Aries and Capricorn inside. It is opposite to the usual way. He writes:

"A circular instrument, which is made of metal, constructed with any arbitrary radius by means of a pair of compasses (*karkaṭa*, it also means Cancer), whose circumference is supported loosely, should be made. Then a horizontal and a vertical straight lines, passing through the centre, should be drawn. The upper half of the circumference should be graduated with degrees of three signs (90^0) on both sides. Two (horizontal) lines should be drawn on one-third-less forty one degrees($40^0 40'$) and twenty five and a twelfth degrees ($25^0 5'$) (above the horizontal line passing through the centre)."[4]

"A pair of diurnal circles which are touching them (two horizontal lines) should be drawn. The intermediate space, between the diurnal circle of Capricorn and Aries, and also (between the circles of Aries and) the circumference, should be divided by degrees of obliquity of the ecliptic along the vertical line. Then the intermediate space between the centre and the circle of Capricorn which is the lowest circle should be divided into sixty six degrees. The sun indeed rotates along a certain circle which is called Cancer etc."[5]

The above mentioned values $40^0 40'$ and $25^0 5'$ show that the obliquity of the ecliptic[6] was taken as about $23^0 50'$, although the value of the co-obliquity of the ecliptic is given as 66^0 in the next verse, and calculation is made by taking the obliquity of the ecliptic as 24^0 which is the common value in Hindu astronomy. The value $\varepsilon = 23^0 50'$ does not appear in the earlier Hindu works, but it is close to the Ptolemy's value[7], $\varepsilon = 23^0 51' 20''$. On the contrary, Mahendra Sūri[8] used $\varepsilon = 23^0 35'$ which is the same as al-Battānī's value[9]. It seems that the measure of the instrument has been borrowed from certain Islamic source, which is different from the source of Mahendra Sūri, but the theory of the instrument is explained in Hindu traditional manner.

The author Padmanābha further continues to explain the method to draw the six o'clock line, prime vertical, altitude circles etc. quoting Śrīdhara, Brahmagupta and Bhāskara II in his commentary.

He wrote that six instruments were described[10], but only two *adhikāras*, which describe one instrument each, are now available.

DIKSĀDHANA-YANTRA OF PADMANĀBHA

The Oriental Institute of Baroda has a unique manuscript of the *Diksādhana-yantra* written by Padmanābha[11]. It consists of 18 verses. D.Pingree conjectured that it is Chapter I of the *Yantra-kiraṇāvalī* or *Yantra-ratnāvalī*[12], but it is wrong because the colophon of the *Yantrarāja-adhikāra* of the Tagore Library (Lucknow University) clearly states that it is Chapter I of the *Yantra-kiraṇāvalī*, hence the *Diksādhana-yantra* cannot be Chapter I of the *Yantra-kiraṇāvalī*. The manuscript of the *Diksādhana-yantra* does not mention the title *Yantra-kiraṇāvalī* nor the *Yantra-ratnāvalī*.

The *diksādhana-yantra* is a wooden horizontal square board with a vertical 12 *aṅgula* gnomon at its centre. A circle of radius 20 *aṅgulas* is drawn at its centre, and concenteric circles are drawn inside at every *aṅgula*. Then east-west and north-south lines, passing through the centre, are drawn. He gives the *agrā* (radius of the circle into sine of amplitude) corresponding to the radius which is equal to the desired hypotenuse (the hypotenuse of a triangle whose base is the desired shadow and upright is the 12 *aṅgula* gnomon) as follows:[13]

$$Agr\bar{a} = \frac{R\sin\delta \times palakar\d{n}a \times i\d{s}\d{t}akar\d{n}a}{R \times 12}$$

where R is the radius of the celestial sphere, δ is the declination of the sun, *palakarṇa* is the equinoctial midday hypotenuse (i.e. (12/cos ϕ, where ϕ is the terrestrial latitude), and *iṣṭakarṇa* is the desired hypotenuse. This *agrā* means the difference between the length of the equinoctial midday shadow and the north-south projection (*bhuja* or base) of the desired shadow. He instructs to obtain east-west projection (*koṭi* or upright) of the shadow from the *bhuja* applying the Pythagorean theorem. He requires to find time using *bhuja*, but the method is not explicitly given. He also asks to draw the locus of the tip of the shadow. The locus is considered to be a circle which passes through the tip of the midday shadow and the tips of the shadows whose corresponding hypotenuse is 60 *aṅgulas* in the morning and evening. He writes:

"The north-south projection (*bāhu* or *bhuja*) of the shadow and the east-west projection (*koṭi*) of the shadow which are stated before should be determined from the 60 *aṅgula* hypotenuse of shadow. The north-south projection should be diminished by the midday shadow. It is the arrow (versed sine). The *koṭi* is the desired sine. Determine the measure of the circle with the help of them and midday shadow. If the circle is drawn with that diameter, then the tip of the shadow of the desired gnomon will not leave its circumference on that day"[14].

PRATODYA-YANTRA OF GAṆEŚA-DAIVAJÑA

Gaṇeśa Daivajña (b. AD 1507) wrote the *Pratoda-yantra*[15], which consists of 13 verses. It is a kind of sun-dial with a horizontal gnomon.

Munīsvara (b. AD 1603) described this instrument in his *Siddhānta-sārvabhauma*[16] in 8 verses[17]. It has a commentary by Munīsvara himself. The extract of this Munīsvara's version was frequently copied. Munīsvara calls it *pratoda-yantra* in the text[18], but calls it *cābuka-yantra* in the commentrary. Hence the extract is sometimes entitled *Cābuka-yantra*[19] and sometimes *Pratoda-yantra*[20]. In the case of the latter, the name of the instrument in the comentary was changed into *pratoda-yantra*. The Munīsvara's version has been published[21]. Sometimes this Munīsvara's version is wrongly stated as *Gaṇeśa's* work. D.Pingree mentions a *ṭīkā* by *Gaṇeśa* himself on the *Pratoda-yantra*[22], but its existence is doubtful[23].

CONCLUSION

The history of astronomical instruments in India between Bhāskara II and Jai Singh Sawai is still unclear although there are several sources. The present paper is only a preliminary report of this subject on which I am now doing research.

ACKNOWLEDGEMENTS

I am thankful to Dr. K. S. Shukla, Retired Professor in Mathematics, Lucknow University, who is guiding my research.

I am also grateful to the Directors and/or Librarians of the following Libraries who kindly allowed me to consult manuscripts (The abbreviations used in the notes are indicated within brackets).

1 Tagore Library, Lucknow University, Lucknow(Lucknow).
2 Sarasvati Bhavan, Sampurnanand Sanskrit Vishvavidyalaya, Varanasi (Benares).
3 Vishveshvaranand Vishva Bandhu Institute of Sanskrit and Indological Studies, Panjab University, Hoshiarpur (VVRI).
4 Scindia Oriental Research Instiute, Vikram University, Ujjain (SOI).
5 Oriental Institute, Baroda (Baroda).
6 The Asiatic Society of Bombay, Bombay (AS Bombay).

REFERENCES AND NOTES

1 Dikshit,S.B. (1981). Bharatiya Jyotish Shastra, Part II. English tr. by R.V. Vaidya, p.231. India Meteorological Department. Garrett,A.ff.(1902). The Jaipur Observatory and its Builder, pp.62-63. Allahabad. I am grateful to the Librarian of BHU who supplied me its photocopy. There are several manuscripts of this adhikāra. See Pingree,D. (1981). Census of the Exact Sciences in Sanskrit, Ser.A, vol.4, pp.170-172. Philadelphia: American Philosophical Society. (hereafter Census). I have used VVRI 2481 and 469; AS Bombay 2451 (BD 298); and Baroda 9588 and 3168. In VVRI 2481, the name of the author is wrongly indicated as Gaṇeśa-Daivajña.

2 Lucknow 45888, 33 ff,copied in Saṁvat 1634 Mārgaśīrṣa-month śukla-pakṣa 8th tithi Monday (= AD 1577). Its colophon clearly states that it is Chapter I of the Yantrakiraṇavalī. (Srīpadmanābhaviracitāyāṁ yantrakiraṇāvalyāṁ yantrarājādhikāro vāsanābhāṣyasahitaḥ prathamaḥ).

3 Although the name of the commentator is not given in its colophon, there is a cancelled colophon in the folio 21b, which states that commented by himself (svavivṛtti).

4 Verse No.3. (Folio 3a).

5 Verse No.4. (Folio 3a-3b).

6 As was shown by Padmanābha in the verse No.6 (Folio 7b), the radii of the diurnal circles of Aries and Capricorn are:

$$a = \frac{r \times R\sin(90^0 - \varepsilon)}{R + R\sin\varepsilon}, \qquad b = \frac{r \times R\text{versed}\sin(90^0 - \varepsilon)}{R + R\sin\varepsilon}$$

where r is the radius of the instrument, a and b are the radii of the circles of Aries and Capricorn respectively, ε is the obliquity of the ecliptic. Therefore, the following equations give the value of ε which gives the values mentioned in the verse No.3.

$$40^0\ 40' = \sin^{-1} \frac{\cos \varepsilon}{1 + \sin \varepsilon}, \text{ and}$$

$$25^0\ 5' = \sin^{-1} \frac{1 - \sin \varepsilon}{1 + \sin \varepsilon}$$

The former gives $\varepsilon = 23^0 49'8''$ and the latter gives $\varepsilon = 23^0\ 51'48''$.

7 Neugebauer, O. (1975). A History of Ancient Mathematical Astronomy, Part I. p.31. Berlin Heidelberg New York: Springer Verlag.

8 Dikshit,S.B. op.cit., p.231.

9 Kaye,G.R. (1918). The Astronomical Observatories of Jai Singh, p.136. Calcutta: Archaeological Survey of India.

10 Commentary on verse No.3.

11 Baroda 3160, 2 ff, copied in Saṁvat 1639 Mārga-month 15th tithi Thursday (= AD 1582).

12 Pingree, D. Census, A-4, p.170 and also Pingree, D.(1981) Jyotiḥśāstra, A History of Indian Literature VI-4. p.53 Wiesbaden: Otto Harrassowitz.

13 Verse No.4.

14 Verse No.12 & 13.

15 I have used Benares 35702 and AS Bombay 245 IV (BD 298). Benares 35702 contains Munīsvara's version also.

16 Although its Pūrvārdha has been published in 3 volumes from Sampurnanand Sanskrit Vishvavidyalaya, Varanasi, its Uttarārdha which contains Yantra-adhiyāya is yet unpublished. I have used Benares 36922 and SOI 9421 (these two are text only), Baroda 9429 and AS Bombay 288 (BD 62) (these two have auto-commentary).

17 Verse Nos.63-70.

18 Verse No.63.

19 I have seen Benares 34999.

20 I have seen Benares 36676, 35630, 34353 and 35074; Baroda 3190; SOI 9414 and VVRI 4731.

21 Sharma,S.D. (1982). Pratoda Yantra. (Ed. and com. by Shakti Dhara Sharma), P.O.Kurali(Ropar) Pb. (INDIA): Martand Bhavan. I am grateful to Dr. S.D.Sharma, Dept. of Physics, Punjabi University, Patiala, who kindly gave me a copy of his book.

22 Pingree, D.(1971) Census, A-2, p.106. Philadelphia.

23 Among 10 manuscripts which Pingree mentions, I have confirmed that the following manuscripts are Munīsvara's version. Benares 36676, 35630 and 34353; Baroda 3190; and VVRI 4731.

DISCUSSION

S.M.R. Ansari : In what other instruments do you find Arabic Islamic instruments ?
Have you found any other Sanskrit source in which Arabic-Islamic influence appears ?
For which latitude the plates of Astrolabe constructed ?

Y. Ohashi : I have not found other instruments of Padmanābha. I cannot say definitely at this moment, but I suppose that there are some other Sanskrit sources which describe Islamic instruments. Since this text is still under study, I would like to present those details on another occasion.

Ram Yantra. A cylinder structure, 5 m high and 7 m in diameter for measuring the azimuth and zenith angles.

4.7 MĀDHAVA'S RULE FOR FINDING ANGLE BETWEEN THE ECLIPTIC AND THE HORIZON AND ĀRYABHATA'S KNOWLEDGE OF IT

R.C. Gupta
Birla Institute of Technology, P.O. Mesra, Ranchi 835215
India

INTRODUCTION

In Fig.1, let U be the rising point of the ecliptic *(udaya-lagna)*, T be the nonagesimal *(tribhona-lagna)* and M be the meridian-ecliptic point *(madhya-lagna)*. Since T is at a distance of one quadrant from U along the ecliptic, the complement of the zenith distance of T, that is the arc TK, will be the required angle between the ecliptic UTM and the horizon NUEKS. The equivalent problem in Indian astronomy is therefore to find what is called *'dṛkkṣepa-jyā'* or the sine of the zenith distance of the nonagesimal.

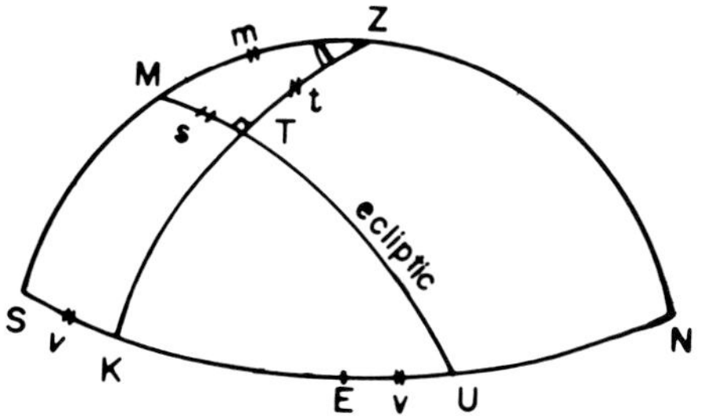

Fig: I

In Fig.1, all angles at T and K are right angles, and it is easily seen that

arc SE = arc KU = arc TU = arc 90^0

so that

arc EU = arc SK = angle MZT = v, say.

If t and m denote the zenith distance of T and M respectively, then the required angle TUK = arc TK = 90-t.

From the spherical right-angled triangle ZMT we have

$$\sin s = \sin m . \sin v \quad (1)$$

$$\text{and } \cos m = \cos s . \cos t \quad (2)$$

where s denotes the arc MT.

Converting all cosines to sines in (2) and solving for sin t, we get

$$\sin t = \sqrt{(\sin^2 m - \sin^2 s) / (1-\sin^2 s)} \qquad (3)$$

while (2) directly gives

$$t = \cos^{-1} (\cos m / \cos s) \qquad (4)$$

Relation (4) may be said to represent the modern solution for finding t from known m and s by using spherical trigonometry (that is, by working on the surface of the celestial sphere). Mathematically it is equivlent to (3).

MĀDHAVA'S RULE

A mathematically correct rule for finding the *dṛkkṣepa* (= arc ZT) as given by Mādhava of Saṅgamagrāma (ca.1460-1425) has been quoted by Nīlakaṇṭha Somayājī (ca. 1500) in his commentary on the *Āryabhaṭīya* of Āryabhata I (born 476 A.D.). The rule is as follows (Pillai 1957, p.75).

Lagnaṃ tribhonaṃ dṛkkṣepalagnaṃ tanmadhyalagnayoḥ/
Vargīkṛtyāntarāla-jyāṃ madhyajyāvargatas-tyajet//
Trijyākṛteśca tanmūle kramaśo guṇahārakau/
Tābhyāṃ dṛkkṣepa-saṃsiddhiḥ trijyāyā jāyate sadā//

"Take the square of the sine of the (arcual) distance between the nonagesimal which is three signs short (in longitude) of the orient-eliptic point and the meridian-ecliptic point. Subtract it from the square of the sine (of the zenith distance) of the meridian-ecliptic point as well as from the square of the radius. The square roots of those two results) are respectively the multiplier and divisor (of the radius). When so operated, the radius will always give the true (sine of the) zenith distance of nonagesimal".

That is, $$\sqrt{(R \sin m)^2 - (R \sin s)^2} = \text{multiplier} \qquad (5)$$

$$\sqrt{R^2 - (R \sin s)^2} = \text{divisor} \qquad (6)$$

Then

$$R. (\text{multiplier}) / (\text{divisor}) = R \sin t \qquad (7)$$

On substitution from (5) and (6) into (7), we find that Mādhava's rule is exactly equivalent to (3), the modern formula employing only sines.

Slightly earlier Nīlakaṇṭha in his *Tantra-saṅgraha* (1500 A.D.) (Sarma 1977), Verses 5-7 had given a similar rule but first computing R sin s by a formula equivalent to (1), But by quoting Mādhava by name, he has now made clear that the real credit for giving the correct rule in explicit form goes to Mādhava. In view of this, the guess of Sengupta (1934) that the correct rule was "perhaps first noticed by Raṅganātha (ca.1603)" is rendered wrong.

However, certain remarks made in the NAB (Pillai 1957) concerning Mādhava's rule show that Nīlakantha is crediting even Āryabhaṭa I for knowing it. We discuss that in the next section.

ĀRYABHAṬA'S RULE

The rule given by Āryabhaṭa I (b, 476 A.D.) in his *Āryabhaṭīya* IV (Gola), 33 for finding the sine of the zenith distance of the nonagesimal (central ecliptic point) is as follows.

Madhyajyodayajīvā-saṃvarge vyāsadalahṛte yat syāt/
Tanmadhyajyākṛtyor-viseṣamūlaṃ svadṛkkṣepaḥ//

Shukla and Sarma (1976, p.144) translated it as

"Divide the product of the *madhyajyā* and the *Udayajyā* by the radius. The square root of the difference between the squares of that (result) and the *madhyajyā* is the (Sun's or Moon's) own dṛkkṣepa".

According to their explanation (Shukla & Sarma 1976 , p.145), the first part of the above rule gives

$$R \sin s = (R \sin m).\ (R \sin v)/R \qquad (8)$$

and the second part then gives

$$R \operatorname{Sin} t = \sqrt{(R \sin m)^2 - (R \sin s)^2} \qquad (9)$$

which, they further say, "is obtained by treating the triangle formed by the Sines of the sides of the spherical triangle ZTM as plane right-angled triangle (which assumption is however incorrect)". Since(9) is only as approximate formula, the above interpretation or translation discredits Āryabhaṭa I for not knowing the correct or exact rule which should be equivalent to (3).

However, the remarks made by Nīlakaṇṭha in his *NAB* about the above rule and the corresponding rule of Mādhava (see Section 2 above) show that Āryabhaṭa knew the correct rule. Just after quoting Mādhava's rule, Nīlakaṇṭha says (Pillai 1957, p.75).

Atra yā dṛkkṣepalagna-madhyalagnāntarālajīvā saiva madhyajyodayajīva-saṃvargad vyasadalāptā,Itah paramubhayatrapi samānaṃ karma.

"Here (in Mādhava's rule) what is called R sin s is the same as the product of R sin m and R sin v divided by R (in the Āryabhaṭa's rule). After this both the procedures are the same (*samānam*)".

That is, the difference between the two rules is only an initial one in the sense that Mādhava takes directly the quantity R sin s which itself is first calculated by Āryabhaṭa by using (8). Then there is no difference. *NAB* (Pillai 1957, p.75) continues and states that what is called "*svadṛkkṣepa*" by Āryabhaṭa is verily the same (*saiva*) as

$$\sqrt{(R \sin m)^2 - (R \sin s)^2}$$

explicitly taken by Mādhava.

In other words, the second part of Āryabhata's rule does not represent the final *dṛkkṣepajya* (= R sin t) as Shukla and Sarma (1976) think but gives only

$$svadṛkkṣepa = \sqrt{(R \sin m)^2 - (R \sin s)^2} \quad (10)$$

instead of (9)

The *NAB* (Pillai 1957, pp.75-76) attaches a special significance to the prefixing word *sva* (literally "own") and explains how to obtain the actual or final *(param) dṛkkṣepa* from the *svadṛkkṣepa* given by (10). Nīlakaṇṭha says (Pillai 1957, p.76).

Yat punariha svasabdena sūcitaṃ trairāsikaṁ tadeva mādhavena vispaṣṭam pradarsitam

"The Rule of Three which is indicated here by the word *sva* (in Āryabhaṭa's rule), the same has been explicitly given by Mādhava(in the last part of his rule)".

That is (with some more details avilable in NAB), Āryabhaṭa's *svadṛkkṣepa* represents only an intermediary step as the sine-chord in a circle of radius R cos s from which the true sine of the zenith distance of the nonagesimal is to be obtained by adjusting the value to the standard circle of radius R by the Rule of Three thereby giving

$$R \sin t = (svadṛkkṣepa) . R/R \cos s \quad (11)$$

On substitution from (10) into (11) we see that Āryabhaṭa's rule will yield the correct value which is same as more clearly expressed by Mādhava. In fact the exposition in *NAB* (Pillai 1957) gives the impression that Mādhava is only elaborating Āryabhaṭa's rule but in more explicit form. If this interpretation is accepted Āryabhaṭa is to be credited for knowing the correct rule and the modern translations of his text in *Āryabhaṭīya*, IV (Gola), 33 are to be modified.

RATIONALE AND CONCLUDING REMARKS

That the rule discussed above was correctly known to Āryabhaṭa is also shown by his knowledge of the correct solution of a mathematically similar problem of finding the right ascension from a given longitude and obliquity (and declination which itself depends on the longitude and obliquity). With reference to Fig.2,

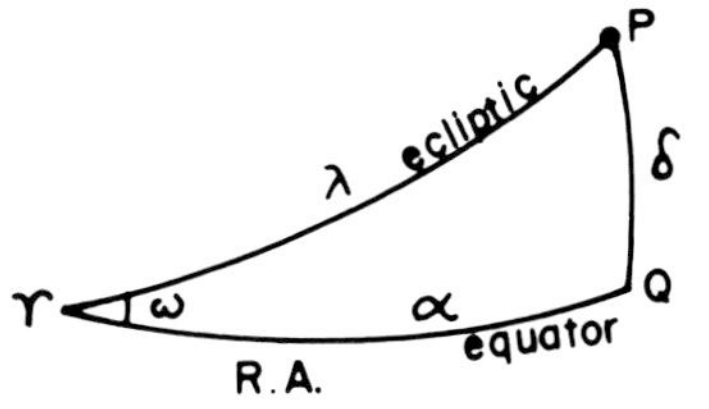

Fig. 2

the solution given in *Āryabhaṭīya*, IV, 25 is equivalent to (Shukla & Sarma 1976, pp.133-134)

$$R \sin\alpha = (R \sin \lambda).\ (R \cos w) \,/\, R \cos \delta \qquad (12)$$

If we apply this formula to the solution of the analogous spherical right-angled triangle ZMT in Fig.1, we get

$$R \sin t = (R \sin m).(R \cos v) \,/\, R \cos s \quad (13)$$

which can be easily seen to be equivalent to

$$R \sin t = \left[\sqrt{(R \sin m)^2 - \left(\frac{(R \sin m).(R \sin v)}{R}\right)^2}\right] (R/R \cos s)$$

$$= \left[\sqrt{(R \sin m)^2 - (R \sin s)^2}\right] (R/R \cos s)$$

by (8). Thus we get the desired rule.

Once we note that the problem of finding *dṛkkṣepa* is exactly analogous to that of finding the right ascension, the derivation of the rule for the former will be similar to that of latter. And the Indian derivation of (12) by working inside the celestial sphere depends on applying the the *trairāsika* (Rule of Three) twice. Details of this simple rationale are already known (Gupta 1974, Shukla & Sarma 1976, p.134).

REFERENCES

Gupta,R.C. (1974). Some Important Indian Mathematical Methods as conceived in Sanskrit Language. Indological Studies, Vol.3 (Pusalker Commemoration Vol.), Nos.1-2, pp.53-55, New Delhi: University of Delhi.

Pillai,S.K. (1957). Edition of The Āryabhaṭīya with the Bhāṣya of Nīlakaṇṭha (= NAB), Part III (Golapāda), p.75. Trivandrum, India: University of Kerala. (This commentary is denoted by the abbreviation NAB in the paper. The quoted rule occurs under Golapāda, stanza 33.

Sarma,K.V. (1977). Edition of The Tantra-saṅgraha of Nīlakaṇṭha. pp.292-293. Hoshiarpur, India: V.V.B.I.S.I.S.

Sengupta,P.C. (1934).Translation of the Khaṇḍakhādyaka of Brahmagupta, p.187. Calcutta,India: University of Calcutta.

Shukla,K.S. & K.V.Sarma (1976). Critical edition & translation of Āryabhaṭīya, New Delhi: Indian National Science Academy.

4.8 BIBLIOGRAPHICAL NOTES ON ISLAMIC ASTRONOMY, THE RESULTS OF A STUDY OF THE EXACT SCIENCES AMONG THE JEWS OF YEMEN

Y. Tzvi Langermann
Institute of Asian and African Studies, Hebrew University, Israel

Introduction

The people of Yemen as a whole, and the Jews of that country in particular, possess a very rich cultural heritage, including achievements in the field of astronomy. The medieval astronomical sources were the subject of an exhaustive study by David A. King (Mathematical Astronomy in Medieval Yemen, Malibu 1983). Some material concerning Jewish interest in the subject was collected by Bernard R. Goldstein (The Survival of Arabic Astronomy in Hebrew, J. for the Hist. of Arabic Science 3, 1979,31-39, note 2c.) The author of this paper has recently completed a monograph on the exact sciences among the Jews of Yemen. In this paper we present some discrete items of mainly bibliographic interest which emerged from that study. Note that our sources are all Arabic manuscripts, written in Hebrew characters.

1 The Zījes of al-Fārisī

Both King (p. 25, no.6.3) and E.S.Kennedy (Survey of Islamic Astronomical Tables, no. 54) report one zīj from Abū ᶜAbd Allāh Muḥammad ibn Abī Bakr al-Fārisī, known by three titles: al-Khāzāʾinī, al-Muẓaffarī, and al-Fārisī. On the basis of certain remarks of Āluʾel ben Yeshaᶜ, a Jewish astronomer working at the very end of the 15th century, we learn that, in fact, the Khazāʾinī zīj and the Muẓaffarī zīj are distinct from one another and different in their makeup. It also appears that al-zīj al-Fārisī is a general term which may be applied to either.

In a discussion of the "second correction" for the five planets, i.e. for the epicyclic diameter at mean distance, Aluʾel writes: "the explanation of this in the Maᶜārij and in the tables of the Khazāʾinī zīj is clearer than that of the Muẓaffarī zīj." (Ms. Heb 28^{0} 6055, Jewish National and University Library, Jerusalem, f58a). The Maᶜārij is another work of al-Fārisī, Maᶜārij al-Fikr al-Wahīj fī ḥall mushkillāt al-zīj (King, no.6.24).

More details as to the differences between the two zījes emerge from the discussion on the equation of time. Aluʾel writes (30a): "It is clear from the al-Fārisī zīj that the extremum of this correction is approximately 30 minutes--so it is in the Muẓaffarī. In the Khazāʾinī it is half of this, and different as well. Up to the present we do not know the reason for this difference."

From this passage we infer that the al-Fārisī zīj may have been a collective title for all the tables of al-Fārisi. (In his commentary to Maimonides' Sanctification of the New Moon, however, Aluʾel speaks of al-zīj al-Fārisī al-Muẓaffarī). More importantly, we learn that the values for the equation of time tabulated in the Muẓaffarī zīj were approximately double those of the Khazāʾinī. Now this raises several problems. First, we note that in his commentary to Maimonides, Aluʾel notes that the lunar corrections found in the Muẓaffarī zīj are double those of the standard zījes. Regarding the second lunar correction, whose maximum is usually about 5^{0}, Aluʾel writes (21a): "The author of the Muẓaffarī zīj doubled it, making it approximately 10^{0}, in the same manner that he doubled the first correction." Aluʾel goes on to say: "Even now we do not know the truth regarding some of the matters included in this zīj, because in it are things not found in the [standard] tables." In fact, however, this doubling of the values is readily understandable, and the suitable explanation was given in an anonymous note to the copy of the Muẓaffarī zīj found in the collection of Rabbi Yosef Kafaḥ of Jerusalem. Speaking of the first lunar correction, whose maximum is about $\pm 13^{0}$, the commentator notes that al-Fārisī subtracted about 13^{0} from all the mean anomalies and doubled the correction, such that the correction would be always positive, and computation simpler.

However, in the case of the equation of time, it is the Muẓaffarī zīj which has the standard values (maximum 30'; cf. O. Neugebauer, A History of Ancient Mathematical Astronomy, 985, 1406), while the Khazāʾinī presents roughly half these values. Moreover, I take the phrase of Aluʾel, "and different as well" (wa-mukhtalifun ayḍan) to mean that the values in the Khazāʾinī zīj are not consistently half those of the Muẓaffarī, i.e. they may have been calculated in an independent fashion. Finally, we note that Aluʾel has not simply mixed up the two zījes: the same Muẓaffarī zīj which has doubled the lunar corrections has also the normal values for the equation of time (e.g. the copy found in BL Or. 4104).

2 Quṭb al-Dīn al-Shīrāzī (?)

Did the writings of the "Marāgha school", with their tremendous innovations, reach Yemen? There is strong, and, to my mind, convincing evidence in the commentaries of Aluʾel that one such work, the Nihāyat al-Idrāk of Quṭb al-Dīn al-Shīrāzī, was in fact known to Yemeni astronomers. Aluʾel refers some seven times to an astronomer by the name of al-Shirwānī. Three important points of detail argue for the identification of al-Shirwānī with al-Shīrāzī, and this despite the fact that the name al-Shirwānī is known in the history of Arabic astronomy, and, in particular, it was also the name of al-Fahhād who, in turn, was an important source for al-Fārisī. The three points are the following:

1) The full name of the astronomer. In an interesting passage Aluʾel writes (71a): "It has been said that al-Shirwānī is the author of the Tabṣirah, but it is most likely that this is incorrect ... the name of the author of the Tabṣirah ... is ᶜAbd al-Jabbār al-Kharaqī, but the name of al-Shirwānī is Maḥmūd bin Masᶜūd." Now Maḥmūd bin Masᶜūd is part of the full name of al-Shīrāzī. Moreover, we learn from this passage that there was some confusion regarding al-Shirwānī, a fact which may help explain what is, in our opinion, the corruption of the name al-Shīrāzī.

2) The title of the work: In his commentary to Maimonides, Aluʾel gives the full title of "the book of al-Shirwānī" as Nihāyat al-Idrāk fī ᶜilm al-Aflāk. (20b)

There is no such work ascribed to al-Fahhād. However, the book of al-Shīrāzī is called Nihāyat al-Idrāk fī Dirāyat al-Aflāk.

3) The theory. In the passage cited above, where Aluᵓel shows that al-Shirwānī is not the author of the Tabṣirah, we read: "Moreover, al-Shirwānī holds that the sun has an epicycle, but the author of the Tabṣirah is not of that opinion." In another comment (33b), Aluᵓel notes that al-Shirwānī assumes two epicycles in the theories of Venus and Mercury. Both of these details are appropriate to the "Marāgha school."

(Note: I do not have a copy of al-Shīrāzī's work. I sent a passage quoted by Aluᵓel from al-Shirwānī to Dr. George Saliba. Dr. Saliba could not find that exact passage in al-Shīrāzī, but neither he nor I regard this as conclusive).

3 Others

In the private collection of Mr. Yehudah Levi Nahum (Ḥolon, Israel), which will surely prove to be of great value once the very numerous fragments have been identified and/or catalogued, are four pages belonging to the astronomical treatise of Qāsim bin Muṭarraf, composed 319 H. in Cordova. The identification is secured by the title of chapter 12 which is preserved in the fragment and matches that given by Sezgin, vol. 6, 158. The city of Cordova is mentioned as well, and the fragment breaks off "in the year 300 of ... ". The Istanbul manuscript, from which Sezgin (by way of an article by F. Rosenthal) learned of the treatise, contains the unique copy of Qāsim's treatise. It is interesting that such an early Andalusian treatise reached Yemen.

The opening page of a treatise on twilight is found in one of the manuscripts in the collection of Rabbi Kafah. Unfortunately, the page is damaged, and it is impossible to make out either the name of the author or the title of the treatise. Reference is made to works on the same subject by Ibn Muᶜādh (published by B. R. Goldstein in Archive for History of Exact Sciences, 1977) and by another jurist, ᶜAbd al-Raḥmān bin Ṭāhir.

A copy of the Zīj al-Jāmiᶜ purports, according to a somewhat unclear note, to have been copied from Kushyār's autograph, which also contained autograph criticisms and corrections on the part of Bahrām ibn Binyāmin. However, this copy is missing part III of Kushyār's zīj.

Several short quotations from Abū-l-ᶜUqūl are found in a manuscript of Rabbi Kafah, but I do not know if these are taken from any of the works listed by King (pp.25ff.). They deal with the (1) size and distance of the sun and the moon, (2) lunar eclipses, (3) musical ratios of the orbs, and (4) circumference of the earth.

Also worth mentioning are (1) a short fragment from Ibn Yunus' Zīj al-Ḥākimī on the elevation of the pole of the ecliptic and (2) a table from the zīj of Yaḥya ibn Abī Manṣūr.

DISCUSSION

S.M.R.Ansari : Did you find any work on Instruments in Yemen ? If I understand correctly there is extent Zīj-i-Safiha of Al-Khazīnī in Yemen.

Y.Tzvi Langerman : There is some mention of instruments but nothing special.

Sorry - I am referring to the Zīj with the title of al-Khazā'inī by the astronomer Abū Bakr al-Fārisī.

4.9 LA THEORIE ASTRONOMIQUE SELON JABIR IBN AFLAH (English Abstract)

H. Hugonnard-Roche
Centre National de la Recherche Scientifique
9 rue Spontini
75116 Paris, France.

Ptolemy's Almagest has been criticized by Islamic astronomers in two different ways: criticisms of Ptolemaic parameters and criticisms, like those of the Maragha school, of the geometrical models used as they contradicted certain basic principles like the principle of uniform motion.

Jabir ibn Aflah's Islah al-Majisti seems to be outside the two aforementioned ways of criticizing the Almagest for he gives an excellent and faithful qualitative account of his kinematical models. He pretends, on one side, to give a complement to the mathematical basis of the Almagest and, on the other, he seems to consider this work as having, in his time, only a theoretical value; due to the modifications of parameters introduced by Islamic astronomers, the Almagest itself has lost, in Jabir's opinion, all practical value for computation. He also gives a list of mistakes he thinks Ptolemy made, but I will limit myself here to analyse what Jabir seems to consider a methodological lack of which I will give only one example: Jabir criticizes Ptolemy's determination of the relative positions of the centre of the equant, the centre of the deferent and the centre of the universe in the case of the models for the superior planets, for he considers that Ptolemy did not give any proof of the fact that the centre of the deferent is the midpoint between the two others.

Let us remember that, for the superior planets, Ptolemy uses an iteration method in which his starting point is to consider that the centre of the equant and the centre of the deferent are the same point. Jabir criticizes strongly this method and compares Ptolemy to a man who cannot see well and walks backwards and forwards in the middle of a forest. Our author proposes a new method which starts by the determination of the position of the planet's apogee: for that purpose he takes two pairs of positions of the planet on each side of the apsidal line, separated by the same interval of time. These positions are, of course, in opposition to the mean sun. Given the symmetrical character of the planet's movement in respect to the apsidal line, one can obtain immediately the direction of the apogee halving the arc between two symmetrical positions taken from each of the two pairs of oppositions. Once he has determined the position of the apogee, he shows the way to find, independently, the value of the two eccentricities, that is to say, on one part, the distance between the centre of the equant and the centre of the world and, on the other, the distance between the centre of the world and the centre of the deferent.

Ptolemy's method is a remarkable example of the procedure consisting in "saving the phenomena", and is considered by Jabir as being purely approximate whilst he thinks his own method is demonstrative. He is probably referring, in an implicit way, to Aristotle's theory of demonstration when he rejects Ptolemy's approach because it is based on the postulate of the bisection of eccentricity, on one side, and on the false supposition that the centre of the deferent and the centre of the equant are the same point, on the other. We should finally say that Jabir's method, even if it is methodologically correct, becomes difficult when put into practice for we cannot observe easily two pairs of oppositions satisfying Jabir's conditions. The necessary time to find these four observations might imply a change in the position of the apogee.

4.10 TRIGONOMETRY IN TWO SIXTEENTH CENTURY WORKS; THE DE REVOLUTIONIBUS ORBIUM COELESTIUM AND THE SIDRA AL-MUNTAHĀ

Sevim Tekeli
Dil ve Tarih-Cografya Fakültesi, Türk Kültürünü Araştırma Enstitüsü, Ankara, Turkey

In Greece, Autolycos (4th cent. B.C.), Aristarchos of Samos (3rd cent.B.C.), Hipparchos (2nd cent.B.C.), Menelaos (1st cent. A.D.), and Ptolemaos (2nd cent. A.D.) are the forerunners of trigonometry. The Greeks used chords and prepared a table of chords.

Later, the Hindus produced Siddhāntas (4th cent.A.D.). The most important feature of these works is the use of jyā instead of chords, and utkramajyā (versed sine).

In Islam, al-Battānī al-Ṣābī (858-929) used the sine, cosine, tangent, and cotangent with clear consciousness of their individual characteristics.

As is known, trigonometry developed as a branch of astronomy. Although in the thirteenth century Naṣīr al-Dīn al-Ṭūsī (in the Islamic world) and in the fifteenth century Regiomontanus (in the West) established trigonometry as a science independent of astronomy, the essential situation did not change, and the subject went on developing as before.

As we come to the sixteeenth century, Copernicus complete some of the work left unfinished by Regiomontanus, in his famous book De Revolutionibus Orbium Coelestium. Later, the chapter devoted to trigonometry was published separately by this pupil Rhaeticus.

On the other hand, the first book of the Sidrat al-Muntahā, by Taqī al-Dīn of Istanbul, was devoted to trigonometry.

The purpose of my work is to make a comparison between these two chapters, and to show what Taqī al-Dīn accomplished with trigonometry in the sixteenth century.

Copernicus in Book 1, Section 12 of Revolutionibus divided the circle into 360^0 and the diameter into 200000 parts.

Taqī al-Dīn in Book 1, Section 1 divided the circle into 360^0 and the diameter into 120 or 2. The first geometrician to divide the diameter into 2 was Abū'l-Wafa᾿ (940-998). In the West, the first mathematician to adopt the simpler form

$$X = 4/3\,X^3 - 1/3\,AD \quad ,$$

where $$X = \sin 1^0, \quad AD = \sin 3^0 \quad .$$

As is seen, Taqī al-Dīn puts everything in terms of the sine or cosine. By contrast, Copernicus mentions only the halves of chords subtending twice the arc. Of course, half the chord of twice the arc is the sine.

On Plane Triangles

In Book 1, Section 13, Copernicus says: Let there be the triangle ABC about which a circle is circumscribed. Therefore arcs AB, BC and CA will be given.

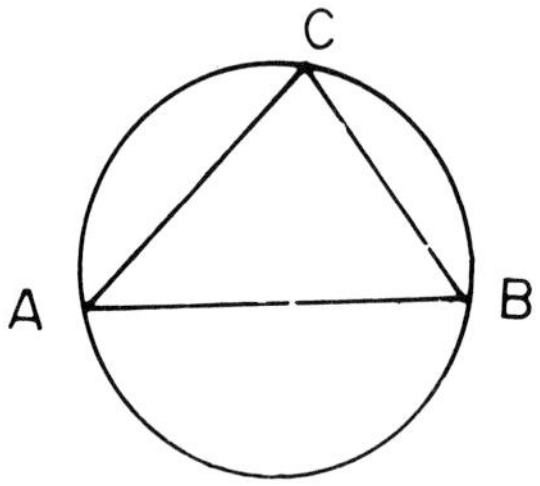

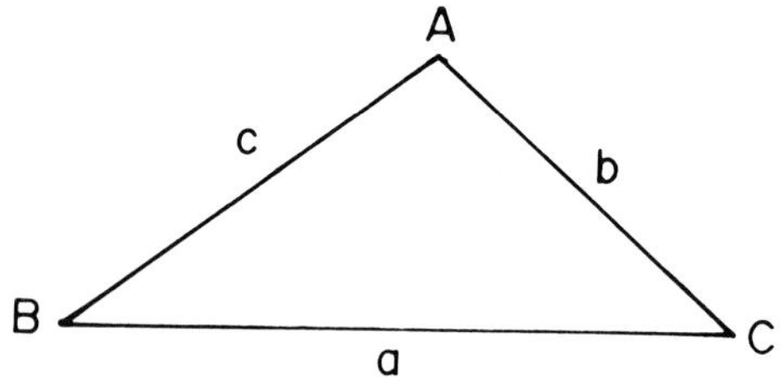

He proves that
(1) if the sides of a triangle, or
(2) two sides and an angle, or
(3) a side and two angles, are given, the triangle is known.

Taqī al-Dīn says in Book 1, Section 4 that the sides of a triangle (it may be acute, obtuse, or right) are proportional to the sines of the angles subtending the sides.

$$\frac{AB}{AC} = \frac{\sin C}{\sin A} \qquad \text{Sine Theorem}$$

As is seen, Copernicus, in solving plane triangles, repeats very nearly what Euclid said in his Elements.

On the other hand, Taqī al-Dīn mentions the important relation known as the sine theorem.

$$\frac{\sin A}{a} = \frac{\sin B}{b} = \frac{\sin C}{c}$$

While the theorem was recognised by Abū'l-Wafā᾽ (940-998) and al-Bīrūnī (973-1048), Naṣīr al-Dīn al-Ṭūsī (1201-1274) was the first to set it forth with clarity.

On Spherical Triangles

Copernicus defines and gives in Book 1, Section 14 the principal properties of a spherical triangle, that is to say, sides neither equal to or greater than the halves of great circles.

$$\sin 90^{\circ} = 1$$

that is to say

$$\text{diameter} = 2$$

was Just Bürgi in the seventeenth century. This is an important point in Taqī al-Dīn's favour.

In the same section Copernicus gives the following formulas for finding crd 2A, crd A/2, crd (A-B), crd (A-B). He also shows that

$$\text{arc } BC / \text{ arc } AB > \text{crd } BC / \text{ crd } AB$$

Copernicus calculated crd 1° or crd 2° by proving that the arc is always greater than the straight line subtending it, but in going from greater to lesser sections of the circle, the inequality approaches equality so that finally the circular lines go out of existence simultaneously at the point of tangency on the circle. Therefore it is necessary that just before that moment they differ from one another by no discernible difference. He established crd 1° as subtended by 1745 approximately.

Taqī al-Dīn also gives the same formulas for crd 2A, crd A/2, crd (A-B), crd(A-B), and he proves that

$$\text{arc } BC / \text{ arc } AB > \text{crd } BC / \text{ crd } AB$$

As we come to Taqī al-Dīn, he says the following, "The Ancients could not find a correct way to obtain crd 1° or crd 2°, in consequence of this, they depended upon approximate methods which do not merit description. " The late Ulug Beg said that we had an inspiration about extracting crd 1° and sin 1°. This method involves the solution of a cubic equation of the form

$$X = \frac{X - AD}{3} ,$$

where X = crd 2° and AD = crd 6°. The leading position of Taqī al-Dīn is obvious.

On Sines

At the end of the 12th section, Copernicus, without using the concept of sine, says, "Nevertheless I think it will be enough if in the table we give only the halves of the chords subtending twice the arc, whereby we may concisely comprehend in the quadrant what used necessarily to be spread out over a semicircle, and especially because the halves come more frequently into use in demonstration and calculation than the whole chords do ".

In Book 1, Section 2 Taqī al-Dīn defines the sine, cosine, and versed sine, and gives the formulas for sin 2A, sin A/2, sin(A-B), sin(A+B).

He says that the radius is greater than the sine, and in the limit the sine of 90° is equal to the radius, that is to say

$$\sin 90^{\circ} = 1$$

Later he calculates sin 1° by using a third equation,

In a spherical triangle having a right angle,

$$\frac{1/2 \text{ crd } 2\ \overline{AB}}{1/2 \text{ crd } 2\ \overline{BC}} = \frac{\text{radius}}{1/2 \text{ crd } 2C}$$

That is to say,

$$\frac{\sin \overline{AB}}{\sin \overline{BC}} = \frac{\text{radius}}{\sin C} \quad .$$

Beside this, in right spherical triangles

(1) if a side and an acute angle, or
(2) three sides, or
(3) two acute angles,

are given, the triangles are known.

Later he mentions the equality of spherical triangles.

In Book 1, Section 5 and a part of Section 6 Taqi al-Dīn also defines and gives the principal properties of the spherical triangle, the sides being neither equal to nor greater than half a great circle.

Spherical triangle ABC having a right angle B.

$$\frac{\sin A}{\sin \overline{BC}} = \frac{\sin 90^{\circ}}{\sin \overline{CA}} \qquad \text{Muġnī theorem}$$

Taqī al-Dīn attributes the Muġnī and its conclusion to Abū Naṣr ibn ᶜIrāq (10th cent.).

1. The first conclusion of the Muġnī:

$$\frac{\cos \overline{YR}}{\cos \overline{BR}} = \frac{\sin 90}{\cos \overline{BY}} \qquad \text{triangle RYB}$$

2. The second conclusion of the Muġnī:

$$\frac{\cos B}{\cos \overline{YR}} = \frac{\sin R}{\sin 90^{\circ}}$$

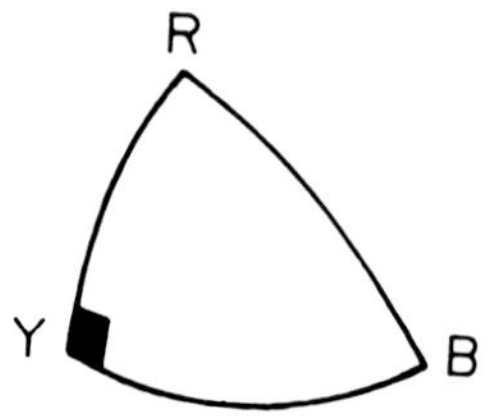

3. The third conclusion of the Muġnī:

$$\frac{\sin RY}{\sin BY} = \frac{\sin CR}{\sin CH}$$

The sides corresponding to equal angles are proportional.

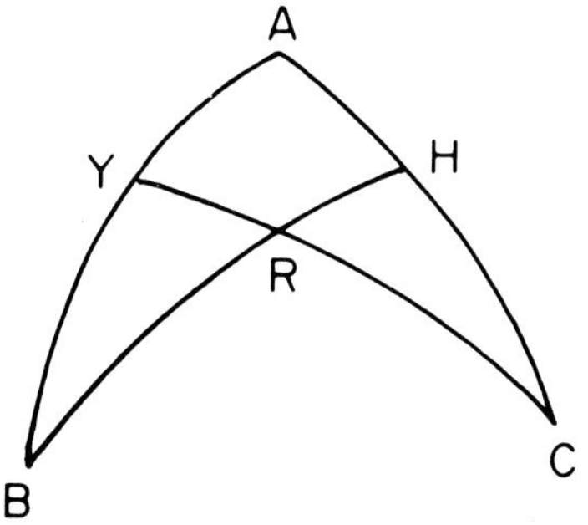

In Section 6, in any triangle having obtuse or acute angles, he gives

$$\frac{\sin AB}{\sin BC} \quad \frac{\sin C}{\sin A} \qquad \text{sine theorem}$$

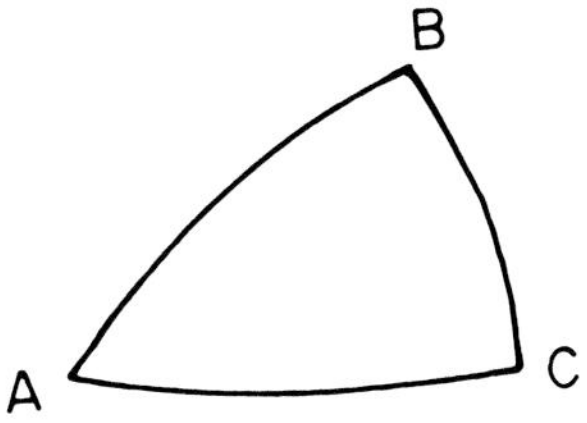

As is seen, there is not only a quantitative, but also a qualitative difference betweeen Copernicus and Taqī al-Dīn who used the Muġnī, the conclusions of the Muġnī and the sine theorem.

On the Tangent and Cotangent
Copernicus says nothing on this subject.

In Book 1, Section 6 Taqī al-Dīn defines the Umbra versa (tangent) and the Umbra recta (cotangent), and gives the following formulas:

$$\frac{\text{tg } A}{\text{ctg } A} = \frac{\text{radius}}{\cos A} \qquad \text{tg } A \text{ ctg } A = \text{radius}^2 \text{ if radius} = 1$$
$$\text{tg } A \text{ ctg } A = 1$$

He gives the tangent theorem and its conclusions.

$$\frac{\text{tg } A}{\text{tg } BC} = \frac{\sin A}{\sin AB} \qquad \text{tangent theorem}$$

1. The first conclusion of the tangent theorem:

$$\frac{\cos B}{\sin 90^{\circ}} = \frac{\mathrm{ctg}\ BR}{\mathrm{ctg}\ BY}$$

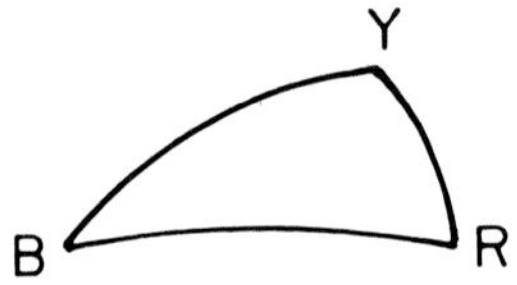

2. The second conclusion of the tangent theorem: (in the same triangle)

$$\frac{\cos BR}{\sin 90^{\circ}} = \frac{\mathrm{ctg}\ B}{\mathrm{tg}\ R}$$

3. The third conclusion of the tangent theorem:

$$\frac{\sin RY}{\mathrm{tg}\ YB} \qquad \frac{\sin RH}{\mathrm{tg}\ HC}$$

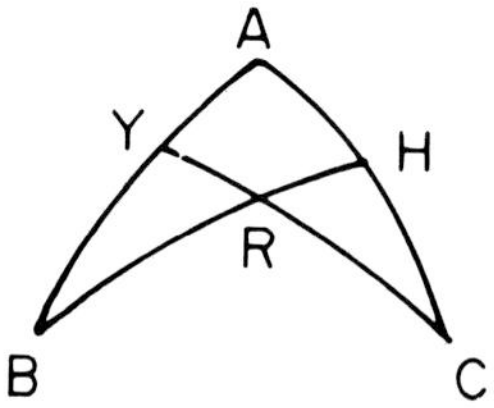

While Taqī al-Dīn presents almost all the arguments for the tangents and cotangents, Copernicus says nothing on this subject.

Conclusion: As is seen, in the sixteenth century, the Islamic World is ahead in the field of trigonometry. Detailed knowledge in trigonometry implies precision in practical astronomy.

4.11 TWO TREATISES ON ASTRONOMICAL INSTRUMENTS BY ᶜABD AL-MUNᶜIM AL-ᶜĀMILĪ & QĀSIM ᶜALĪ AL-QĀYINĪ

S.M. Razaullah Ansari
Department of History of Medicine & Science,
I.H.M.M.R., Hamdard Nagar, New Delhi- 110 062, India.
(Presently at Phys.Dept.,Aligarh Muslim Univ.Aligarh,India)

S.A. Khan Ghori
Department of Physics, Aligarh Muslim University,
Aligarh - 202 001, India.

INTRODUCTION

A characteristic feature of Arab-islamic astronomy during the Middle Ages is the promotion and tremendous growth of practical astronomy which was in turn manifested primarily by the establishment of scores of observatories in West-Central Asia, from Abbasid Caliph al-Māmūn (813-833) to the Turkish king Murād III (1574-1595), and by the production of copious literature on astronomical Tables (the *zījes*) as well as on astronomical instruments *(ālāt al-raṣad)*. The enormity of the literature on the latter could be gauged by the list of extant works as given by Matvievskaya and Rosenfeld (1983) in their recent Bio-bibliography: 349 treatises on astrolabes, 138 on sine-instruments, 81 on quadrants, 4 on sextants and octants, 41 on armillary spheres and celestial globes, 77 on sundials and again 77 on "other instruments"--in all 767 treatises. As a matter of fact the instruments developed by Arab-islamic astronomers could be broadly classified into four groups: a) Time measuring instruments (e.g. sundials, shadow quadrants), b) Angle measuring instruments for astronomical parameters (e.g. armilla of various kinds, dioptre and parallactic rulers), c) instruments for transformation of system of coordinates and/or solving nomographical problems (e.g. astrolabes, quadrants, *dastūr* instrument), d) Mathematical instruments for evaluating trigonometric functions, (e.g. sine-quadrants). Apart from the fourth and the most important of all, the astrolabe, which in turn embodies all the four groups of instruments to a certain extent, works on "other instruments" were compiled in almost every century (down from 9th to 18th A.D.), also by well-known Arab-Islamic astronomer-mathematicians. In order to stress the significance which the Arab-Islamic savants attached to astronomical instruments, we give here the following selected list:

In 9th c., al-Aṣṭurlābī, al-Kindī, Thābit Ibn Qurra, Sulaymān Ibn ᶜIṣma; in 10th c., al-Khāzin, al-Sijzī, al-Khujandī, Ibn al-Haytham; in 11th c., Ibn Sīnā, Abū al-Ḥasan al-Anṣārī; in 12th c., al-Khāzinī, Abū Naṣr al-Maghrabī; in 13th c., al-Marrākushī al-Abharī, al-^{c}Urḍī, ^{c}Alī Shāh al-Khwārizmī, al-Banna; in 14th c., Ibn al-Raqqām, al-Bukhārī, Ībn al-Shāṭir; in 15th c., al-Kāshī, al-Wafā'ī, Sibṭ al-Mārdīnī, Shamsuddīn al-Miṣrī, al-Humāwī; in 16th c., al-Barjandī, Ṭaqī al-Dīn, ^{c}Abdul Mun^{c}im al-^{c}Āmilī, Qāsim ^{c}Alī Qāyinī; in 17th c., Isḥāq Āfindī, al-Humaydī; in 18th c., Mahārājā Sawāi Jai Singh II; in 19th c., Ghulām Ḥusayn Jawnpūrī. For their various treatises one may refer to Matvievskaya & Rosenfeld (1983, Vol.II). However, one finds not many *general* treatises, wholly devoted to *all* astronomical instruments. To our knowledge they are only twelve works, namely by al-Khāzin, Ibn Sīnā, al-Khāzinī, al-Marrākushī, al-^{c}Urḍī, al-Kāshī, al-Barjandī, Taqī al-Dīn, al-^{c}Āmilī, Qāsim ^{c}Alī Qāyinī, Sawāi Jai Singh and Ghulām Ḥusayn Jawnpūrī. The work of Jai Singh with the title *Yantraprakāra* is in Sanskrit though on Arabic instruments.

Following the suggestion of Kennedy (1961) and Seemann (1928) we decided to study the work of al-^{c}Āmilī and later found a similar work in the name of al-Qāyinī. Their unique copies are extant at British Musseum (London) and Mulla Firoze Collection at K.R. Cama Oriental Institute (Bombay) respectively. Our finding is that both monographs are surprisingly *identical verbatim* beginning with the descriptive text of the instruments, the introduction or preface being evidently different, see below. Moreover the monograph of al-^{c}Urḍī (Seeman 1928) seems to be the source of the work in question. We intend to publish the full text with English translation elsewhere.

THE MANUSCRIPTS

The unique Ms of ^{c}Abdul Mun^{c}im al-^{c}Āmilī is extant at British Museum Persian Collection (Per. Add. 7702). It consists of 27 folios, size 8+1/2 x 6 inches, with 23 lines per page. On folio 1b, 6 lines from another beginning of the Ms are given, the last line is incomplete. The text of these lines does not agree exactly with the *actual* another beginning of the Ms which is on folio 2b (see the accompanying photograph 1) folio 2a is blank. It is clear from the photograph that the handwriting of f.2b does not tally with that of f.3a. The remaining folios of the Ms are in the latter handwriting, (also noted by Seeman 1928). The title of the work does not appear in the introductory text; it is noted on the flyleaf or folio 1a: *Kitāb ta^{c}līm ālāt-i* *zīj*. It is a bit unusual, since normally the Islamic astronomers used the term instruments of observation *(ālat-i raṣad)* rather than instruments of *zīj* (the Tables). The name of the author appears at the end of 7th line of f.2b. In the colophon, the year of this copy of the Ms is given as 1112 A.H./1700 A.D.

Not much is known about al-ᶜĀmilī. Suter (1900) did not give any biographical information about him. Matvievskaya & Rosenfeld improved upon Suter just a bit by referring to Sayili (1960) and Storey (1958, p.85). Sayili relied mostly on Seemann who on the basis of the note on the margin of f.2b dated the writing of the Ms as 970 A.H./1562-63 A.D.[1] And therefore Seeman concluded that al-ᶜĀmilī must have lived during the reign of S͟hāh Ṭahmāsp I (1524-1576) - the second safavid king of Iran. However, the author of the manuscript explains in the text of f.2b, line 4 from below, the motivation of the king by the following words: "so that the *riwāq* of the capital of Iṣfahān becomes like that of Alexandria, Marāg͟ha and Samarqand". Now, we may note that S͟hāh Ṭahmāsp's capital was Qazwīn and only during S͟hāh ᶜAbbās I (1587-1629), Iṣfahān became the capital of Safavid Kingdom. To be more precise, ᶜAbbās shifted from Qazwīn to Iṣfahān in the year 1591 (Malcolm 1976; Fisher 1968). In other words, al-ᶜĀmilī must have flourished and wrote, if at all, the tract in question after that year, i.e., at the close of the 16th century or beginning of 17th century.

Again, Qāsim ᶜAlī Qāyinī's manuscript is extant also as a unique copy at Mulla Firoze Collection in K.R. Cama Oriental Research Library, Bombay Ms No.I-21, described for the first time by Rehatsek (1873). The Ms consists of 98 pages, of size 7.4 x 4.9 inches, with 17 lines per page. The title of the work is given by Qāyinī as *Jāmaᶜ al-anwār min al-kawkab wa al-abṣār* on f.1b last line; see photograph 2 and note in another hand writing the heading at the top: *Ṣuwar ālat-i raṣadī*, i.e. diagrams of astronomical instruments. The name of the author is quite clear in the beginning of the 9th line, f.1b. In the colophon, the year is clearly given as 1100 A.H. (i.e. 1689 A.D.), which was, strangely enough, wrongly read by Rehatsek as 1000 A.H., i.e. 1592 A.D. (Storey 1958, p.89; Matvievskaya & Rosenfeld 1983, p.597). This manuscript is in an excellent handwriting, far better than that of al-ᶜĀmilī.

Whereas Rehatsek has said nothing about Qāyinī, Storey gives a couple of biographical details about him, on p.89. On the basis of our own survey of Qāyinī's various works, particularly those extant in the libraries of Soviet Central Asia and Leningrad, we have succeeded in securing a bit more information about him.

Qāsim ᶜAlī was a student of the fairly known astronomer and mathematician Muḥammad Bāqar ibn Zayn al-ᶜĀbidīn al-Yazdī (d. 1637) (Munzavi 1969, p.230, Foreword of MS.Serial No.2024), who was himself a disciple of Bahā'uddīn al-ᶜĀmilī (1547-1622) (Matvievskaya & Rosenfeld 1983, 2, No.490, p.590). Qāyinī stated in the preface of one of his works that he learnt from al-Yazdī the art of making astrolabe.[2] Later, he occupied himself for quite sometime with the construction of astrolabes, sine-quadrant[3] This statement of Qāyinī is borne out by several of his extant tracts on the construction and uses of astrolabes. They are as follows: 1) Ms No.1/699 Sipahsālār Library (Tehran), 40 ff, copied in 1672 by Sayfuddīn Maḥmūd (Munzavi 1969, p.230).. 2) Another Ms at Mawlānā Muḥammad ᶜAlī Library (Atak, Pakistan), 190 ff copied in 17th c. by Muḥ.Ṣāliḥ (Munzavi 1983, p.238, Serial No.1401). 3) One in Mawlānā Qudratullāh Personal Collection (Sargodha, Pakistan), 175 ff, copied in 1776 by Aḥmad ᶜAlī Ḥasan (Munzavi 1983, p.238, Serial No. 1402), 4) Ms No. PNS 114, at State Public Library (Leningrad), 10 ff, ca. 1815-16. 5) Collection No.6400 in Majlis Library (Tehran), 40 ff, copied in 1840 by ᶜAbduljānī ᶜImāduddīn (Haeri 1972). 6) Ms. No.245 in the Library of Centre for Central Asian Studies (Sringar/Kashmir, India), 62 ff, copy of 1841 (Bhatt 1982). We have found this tract to be the only one of Qāyinī's writings which is in Arabic. The title is *Lub-i al-lubāb fī Kayfīya al-ᶜaml bi al-asṭurlāb.* 7) Ms No.4/4061 in Majlis Library (Tehran), copy of 1847(Munzavi 1969, p.230). 8) Also *Risālah Asṭurlāb Zawraqī,* Punjab University Library (Lahore, Pakistan), 9 ff, author's name given therein is Qāsim ᶜAlī (Munzavi 1983, p.240, Serial No. 1420)[4].

Besides the above listed works on astrolabe, Qāsim ᶜAlī also wrote: 9) *Risālah dar maᶜrifat-i Qibla,* Majlis Library (Tehran), Ms. No.2/2377, copied in 1647, Munzavi 1969, p.331. 10) *Risālah dar ᶜIlm-i Hay'at,* Leningrad University Library, Ms. No.402(Matvievskaya & Rosenfeld 1983, 2, No.501a, p.597 entry A1). We have found an autograph copy of a *Risālah Hay'at* in Ms Collection of Ibn Sīnā Library at Bukhara, Ms. No.161 (new), copied in 1059 A.H./1650 A.D. This is a unique Ms with several marginal notes signed by Qāsim ᶜAlī Qāyinī, and has not been cited by Matvievskaya & Rosenfeld (1983). 11) Translation of *Al-jabr wa al-Muqābla* by al-Ṭūsī, Ms No.2/1319 at Tehran University Library, 54 ff, copied in 1671 by Rafīᶜ son of Muhammad Hasan Qāyinī, (Munzavi 1969, p.148). 12) *Tashrīh dar Parqār,* Ms 39 in Mashhad, copied 1656, with marginal notes[5] in Qāyinī's handwriting (Storey 1958, p.89). 13) *Maṭlaᶜ-i Hīlāj,* Ms. No.3/2377 at Majlis Library (Tehran), copied in 1682 (Munzavi 1969, p.351).[6]

To this impressive list of Qāyinī's works *Jāmiᶜ al-Anwār min al-Kawkab al-Abṣār,* copy of 1689, fits quite well as the culmination of Qāyinī efforts for practical astronomy. But before we discuss this presumably last of his work, let us first settle the time when Qāyinī flourished. On the basis of marginal notes in his works No.10 (Bukhara Ms) and No.12 (Mashhad Ms), it is clear that he was alive during 1650-56. On the other hand, Bāqar Yazdī died in 1637 and therefore his pupil at the time of his death must not have been younger than 18 years of age, i.e., born in about 1619 A.D. Therefore when he wrote his first work on *Qibla* about 1645-47 he could be 26 years old and at the time of his last work in 1680-85 he could well be 60-66 years old. So roughly one may say that Qāyinī flourished during 1620-1680.

THE AUTHORSHIP

To settle this question, let us note down the following facts.

Not a single work other than *Kitāb ālāt-i zīj* is known to have been written by ᶜAbdul Munᶜim al-ᶜĀmilī. To write a general work on all astronomical instruments one is expected to be quite well-versed in the state-of-the-art. As listed in the introduction, there are very few general or comprehensive works on instruments. One is therefore justified to be skeptical about al-ᶜĀmilī's authorship. Further we find that in the last line of f.1b, there is an abruptness in the text; see photograph 1. It is stated:

> "...*wa ghalaṭ wa sahl angārī zīj-i mazbūr maᶜlūm ᶜālamyān gardad wa ᶜillat namī twānad ki misṭar ālat dhū shuᶜbatayn gardad*".

That is, "... and the mistakes and omissions or carelessness of the afore-mentioned zij [i.e. Zīj-i Ulugh Beg] could be known to the world and the ruler of triquetrum could not be the reason....." In other words the author abruptly switches on from the theme of previous compiled *Zīj* to the astronomical instrument, the triquetrum. Comparing the texts of the two manuscripts we find that from the words: *misṭar ālat dhū shuᶜbatayn* onwards the two texts coincide *verbatim*, see photographs 1 & 2.

In contradistinction to what we have said above, the case of Qāyinī as an author can be presented as follows. Since Qāyinī had been a student of quite a well-known astronomer Muḥd. Bāqar Yazdī, he became in a short time quite a bit of authority on the construction of a few instruments, like astrolabe and quadrant, in the first place. Further in his writings he has shown keen interest in the practical aspect of the instruments on which he also wrote quite substantially. Therefore it is evident that in the closing years of his life, he could write quite well a monograph on the construction and use of astronomical instruments in general. Moreover one finds in his manuscript references to his various writings which are also extant in various libraries of the world, for instance No.5,9, of our list and Tashkent Ms, (See Note 3). Another interesting fact, which we have found out, is that the title of his monograph differs a wee-bit from the title of the philosophical work of his teacher al-Yazdī namely, *maṭlaᶜ al-anwār wa maṭlaᶜ al-anẓār*, (Matvievskaya & Rosenfeld 1983, 2, No.490, p.590). One may interpret it as a kind of his dedication to the memory of his teacher. Consequently we think that ᶜAbdul Munᶜim just plagarised it by replacing Qāyinī's preface with his own. In fact the first draft of the first para of al-ᶜĀmilī's preface is given on f.1b which is different from what he wrote in the actual manuscript on f. 2b. It appears that al-ᶜĀmilī just copied Qāyinī's manuscript, from where Qāsim ᶜAlī starts talking

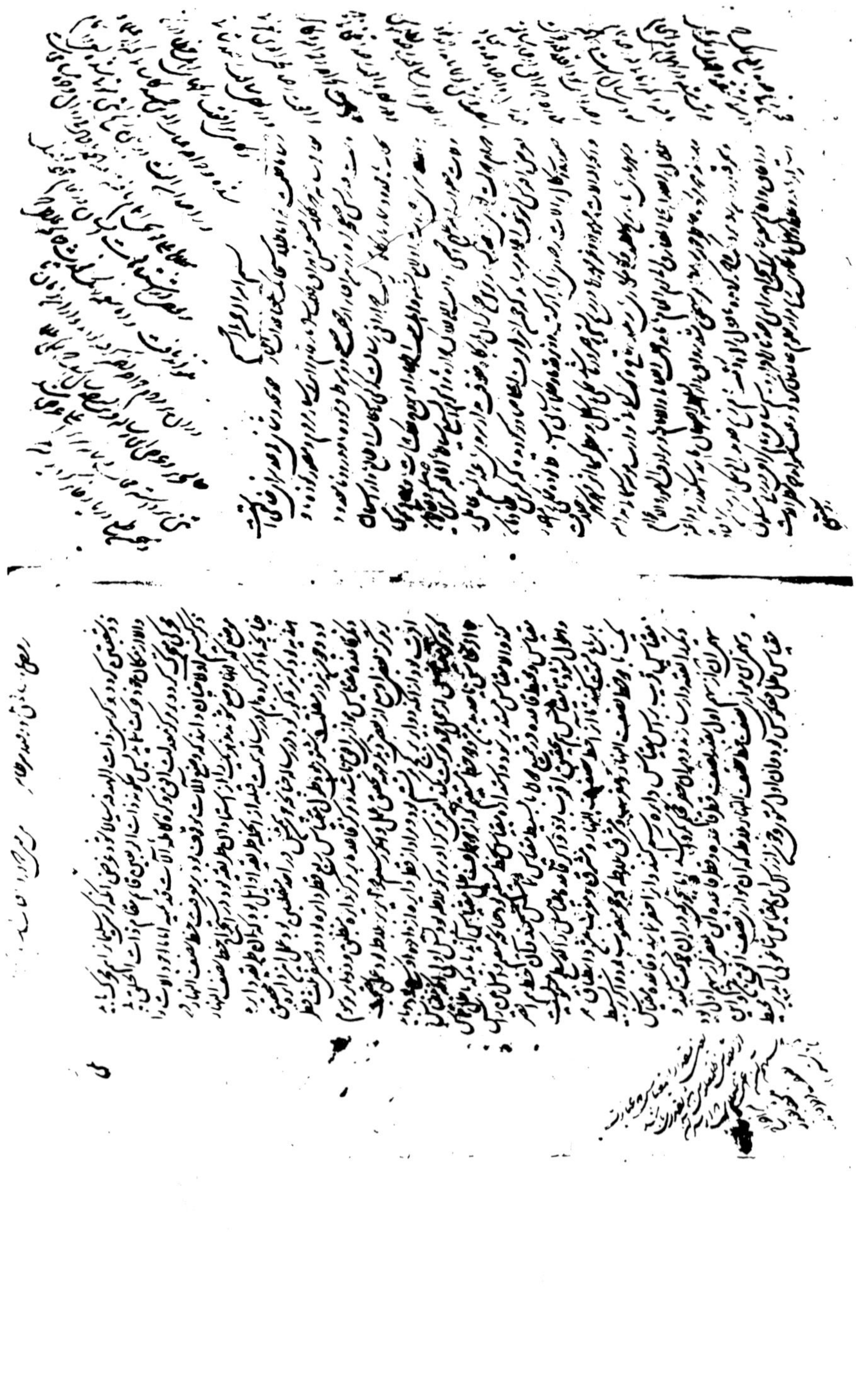

photograph 1 : Folio 2b and 3a of ʿAbdul Munʿim al-ʿĀmilī's *Kitāb ālat-i zīj*, (courtesy British Museum, London).

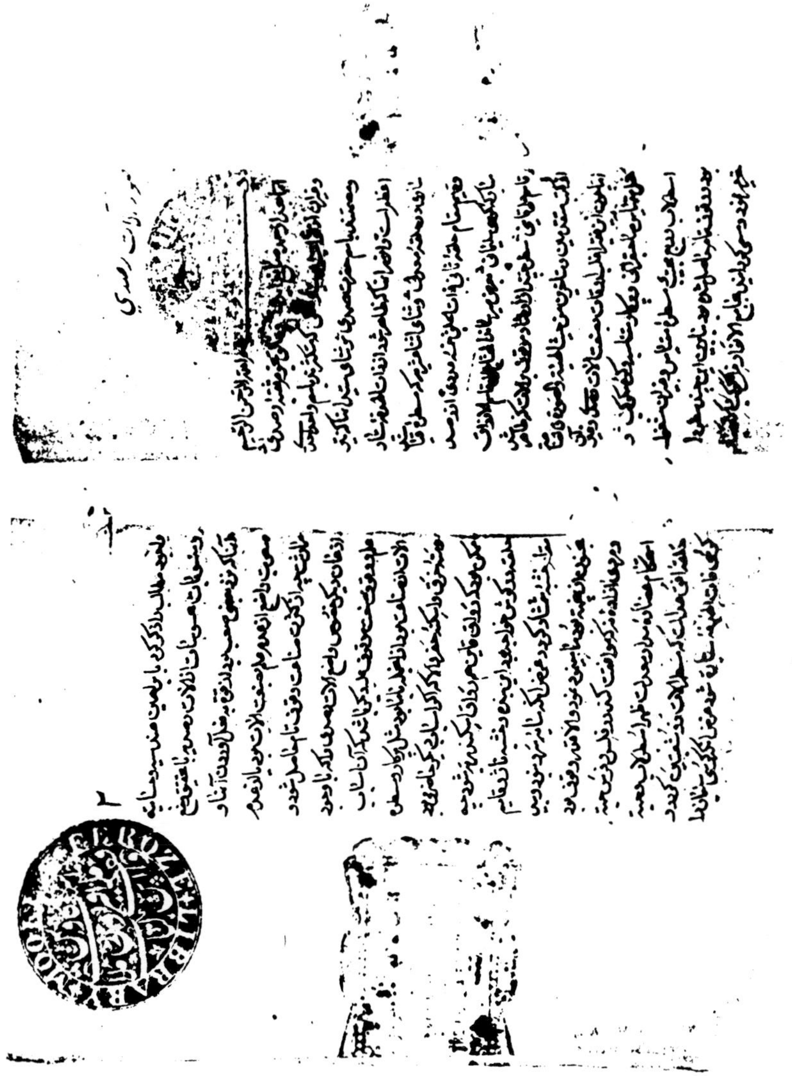

photograph 2 : Folio 1b and 2a of Qāsim ʿAlī Qāyinī's *Jāmaʿ al-anwār min al-kawkab wa al-abṣār* (courtesy K.R. Cama Oriental Institute, Bombay).

about the instruments, though strangely enough without understanding the text. We may add that the text of Qāyinī's manuscript on f.2b starting from the words *misṭar ālat-i dhū shu^c batayn* fits quite well with the sentances which precede these few words. We are therefore of the opinion that the real author of this monograph on astronomical instruments was Qāsim ^c Alī Qāyinī.

CONCLUDING REMARKS

For lack of time and space we confine ourselves only to the following remarks concerning the contents of Qāyinī's tract: Without any definite chapterisation, the work begins after introductory remarks first with an account of the determination of the line of meridian by the method of *Indian circle*. Then follows a detailed description of the construction of various instruments, including methods of installation and of accuracy *(taṣḥīḥ)*. The instruments are: 1) *Dhāt al-rub^c a* (mural quadrant), 2) *Dhāt al-ḥalaq* (armillary sphere), 3) *Ālah-i ma^c rifat-i mayl-i a^c ẓam* (instrument for the determination of obliquity), 4) *Dhāt al-hadaf-i Sayyārah*[7] (dioptre), 5) *Dhāt al-shu^c batayan* (the parallactic rulers or triquetrum)[8], 6) *Dhāt al-rub^c ayayn al-ufuqīyah* (the double quadrants), 7) *Dhāt al-taghyīr* (the modifying instruments), 8) *Dhāt al-sahm wa-l-jayab*, 9) *Dhāt al-sahm wa-l-jayab*, 10) *Ālah-i Kāmilah* (the perfect instrument), 11) Celestial globe, in connection with the account of fixed stars.

Apart from the latter, this list of instruments coincides exactly with that of Mu'ayyad al-Dīn al-^c Urḍī (fl. 13th c.), as given in his tract: *Risālah Kayfīyah al-Arṣād* (Seeman 1928; Tekeli 1970), even the serialization of instruments is identical. Further, Qāyinī's several diagrams of instruments agree closely with those of al-^c Urḍī. But his drawings and perspectives in both manuscripts (i.e., from British Museum and Mulla Firoze libraries) come through better than in al-^c Urḍī's tract.

On comparing Qāyinī's & al-^c Urḍī's tracts, we have found the following mistakes. Qāyinī has wrongly used the word *sahm* in No.8, instead of *samt* as in al-^c Urḍī's corresponding instrument. He exchanged the words *thuqbatyan* (two holes) with *shu^c batayn* (two rulers) and vice versa, (See Notes 7 & 8). Further, whereas Qāyinī's triquetrum is actually the parallactic rulers of Ptolemy, his instrument No.7. is the same as al-^c Urḍī's *dhāt al-usṭuwānatayn*, fig.V in Tekeli's (1970) Arabic text, p.149 (See also Seeman 1928). Another mistake of Qāyinī is regarding the name of the constructor or inventor of the "perfect instrument" (No.10). He attributes it to Najm al-Dīn, *Wazīr* of Malik Manṣūr, ruler of *ḥimṣ*[9]. On the other hand al-^c Urḍī reports: "In the year 650 I constructed *(^c amiltuhu)* another instrument for Malik al-Manṣūr, ruler of *ḥimṣ* in the presence *(biḥuḍūr)* of Wazīr Najm al-Dīn...., (Tekeli 1970, Eng.Tr.p.93, Arabic p.159). In fact we have compared a couple of instruments in the Arabic and Persian texts and have found them to be almost the same. Thus we conclude that the source of Qāyinī is al-^c Urḍī's *Risālah*. As mentioned in the last section Qāyinī quotes in his various works a couple of his predecessors or contemporaries, like his teacher Muḥammad Bāqar Yazdī and his son Mullā Muḥammad Ḥusayn (Munzavi 1969, p.230) , also Muḥ. Taqī bin ʿAbd al-Ḥusayn Naṣīrī (Haeri 1972). It is not clear why he did not quote al-ʿUrḍī as his source. Further light may be thrown on this question and the state-of-the-art of astronomical instrumentation by a detailed study of Qāyinī's work. We intend to publish an English translation with edited text elsewhere.

ACKNOWLEDGEMENT

The authors acknowledge with gratefulness Indian National Science Academy (New Delhi) for sponsoring this work. They are also indebted to British Museum (London) and K.R. Cama Oriental Institute (Bombay) for providing the microfilm and xerox of the Mss respectively. One of us (SMRA) is grateful to Hakeem Abdul Hameed (President of IHMMR, New Delhi) and to Mr. Ausaf Ali (Director of the Institute of Islamic Studies, New Delhi) for providing facilities, also to USSR and Indian Academies of Science for his study tour of the Soviet Central Asia.

NOTES

1 On the margin it is recalled that "due to the precessional motion of the earth, mistakes crept in and can be found in the astronomical Tables of Ulugh Beg, Naṣīruddīn al-Ṭūsī and in the observations of Muḥīuddīn Maghrabī, al-Battānī, Chinese scholars, Jamshīd al-Kāshī." Stressing further that "in it a mistake was attributed to late Khwājā Naṣīr, it is nearly three hundred years from that time to writing and presentation of this treatise, that scholars took the text and presented it... and it got agreed to the deep-ocean like mind". Evidently this marginal remark is concerned with *zījes*, rather than astronomical instruments. And therefore it could not be attributed to al-ʿĀmilī's work. Note that the enumeration of names is not chronological; the scribe was presumably not a scholar! We think that this note is just an *addendum* to the last line of the text (f.2b): ".....and the mistakes and carelessness of the above mentioned zīj [i.e. of Ulugh Beg] could be known by the world". Further the difference of the *hijra* years of writing the Ms and that of the Ulugh Beg's Tables: 1112-842=270 $\approxeq$ 300 years. Then the marginal note will agree well with the above-mentioned text as an additional remark.

2 Ms. PNS, 114 at the State Public Library (Leningrad) with the title *Imtaḥān Asṭurlāb*, see f.1b.

3 Ms. No.465/IV at Institute of Oriental Studies (Tashkent), f.1b. Another incomplete copy Ms.No.5185/IX. This Ms is wrongly entered in the Tashkent Catalogue as on astrolabe.

4 This boat-like astrolabe was invented, in fact, by Abū Saʿīd Aḥmad al-Sijzī (fl.10th c.) as reported by al-Bīrūnī, see Sezgin, 1978. A number of anonymous manuscripts on *Asṭurlāb Zawraqī* are extant: Mālikī Library (Tehran), Raḍā Library

(Rampur), Sālārjang Museum Library (Hyderabad). Mawlānā Dā'ūd Collection (Sargodha, Pakistan), Institute of Manuscripts (Baku, Azarbaijan). The latter is a copy of 1593.

5 He lists also the title *Tashrīḥ al-aᶜmāl* at Mulla Firoze Collection (Bombay) and another copy of 1680 at Qāḍī ᶜUbayd Allāh Library (Madras) under Qāyinī's writings. However, a manuscript *Tashrīḥ al-aᶜmāl dar ᶜamal parkār-i mutanāsibah* by Muḥammad Zamān Usṭurlābī Mashhadī, copied in India in 1862, exists in Raḍā Library (Rampur) No.1163. And therefore Storey conjecture is questionable.

6 Munzavi notes that with this Ms are attached Qāyinī's two other works: *Qibla* and *Maṭlaᶜ al-ḥikam*, the former may be the same as No.9 of our list.

7 Qāyinī calls it also, though wrongly, *dhāt al-shuᶜbatayn* instead of *dhāt al-thuqbatayn*, as it should be.

8 Again he names it wrongly *dhāt al-thuqbatyn*. He calls it also *ālah-i ikhtilāf-i manẓar*.

9 Bombay Ms. f.42b, London Ms. f.23a.

REFERENCES

Bhatt,G.R. (1982). A Handlist of Arabic and Persian Manuscripts of the Research Library, Centre of Central Asian Studies, University of Kashmir, Srinagar.

Fisher,W.B. (1968). The Cambridge History of Iran, ed. **1**, 105-106

Haeri,Abdol Hossain (1972). Catalogue of the Library of Majlis-i Shurā-i Millī, **19**, Tehran, esp. p.555.

Kennedy,E.S. (1961). Al-Kāshi's Treatises on Astronomical Observational Instruments, Journal of the Near Eastern Studies, **20**, 98-108; reprinted in Studies in the Islamic Exact Sciences, ed. D.A. King & Mary Helen Kennedy, Beirut, 1983.

Malcolm,John (1976). The History of Persia, **1**, 506, Tehran.

Matvievskaya,G.P. & Rosenfeld,B.A. (1983). Matematiki i Astronomi Musul'manckovo Srednevekov'ya i ikh Trudi (8-17th Century), 3 vols, Moscow, See esp. **3**, pp.147-152 for astronomical instruments.

Munzavi,Ahmad (1969). A Catalogue of Persian Manuscripts, **1**, Tehran.

- (1983). A Comprehensive Catalogue of Persian Manuscripts in Pakistan, Islamabad, **1**.

Rehatsek,E. (1873). Catalogue raissoné of Arabic, Persian, Hindostani and Turkish Manuscripts, Bombay, esp. pp.14-15.

Sayili,A. (1960). The Observatory in Islam, Turkish Historical Society Series **7**, No.38, Ankara; reprinted Arno Press, New York, 1981, esp. p.288.

Seeman,H.G. (1928). Die Instrumente der Sternwarte zu Marâgha nach der Mitteilungen von Al-'Urḍî , Sitz. Ber. d. physik. Medizin. Soz, zu Erlangen, **60**, 15-126. See esp. his

German translation of al-ᶜĀ-mili's preface, pp. 121-126; For English and Turkish tr. along with Arabic text See Tekeli (1970).

Sezgin,F. (1978). Geschichte des arabischen Schrifttums, **6**,Leiden, p. 224.

Storey,C.A. (1958). Persian Literature,A Bibliographical Survey, Luzac & Co. London, 3 vols, in particular **2**, Pt. I.

Suter,H. (1900). Die Mathematiker und Astronomen der Araber und Ihre Werke, Abhandlungen zur Geschichte der Mathematischen Wissen-chaften, Bd. X, published separately, from Leipzig. Reprinted by APA - Oriental Press Amsterdam. For al-ᶜĀmili, see entry No. 473, p.192.

Tekeli,S. (1970). Al-Urdî'nin "Risalet-ün fī Keyfiyet-il-ersad" Adli Makalesi, Araştirma,**8**, also issued separately, Ankara 1972, pp.169, Edited Arabic text with English and Turkish translation.

DISCUSSION

L.C. Jain : Have you included in your studies the works, 'Yantrarāja' and 'Yantrasiromani', probably of 13th and 15th Century A.D. respectively.

S.M.R. Ansari : I know the works; but I am not concerned at all with astrolabes.

S.Tekeli : As far as we know there is only one sextant described by Taqī al-Dīn. Can you call *suds-i Fakhrī* as a sextant?

S.M.R. Ansari : It is a question of definition, but the word *suds* means 1/6th = 60^0. Isn't it ?

S.N.Sen : Is there any instrument like the Rāma Yantra found in Jai Singh's Observatory in the instrument list of Marāgha Observatory ?

It is known that a few Chinese instruments makers worked at Marāgha under Nasīr al-Dīn's direction. Was the Chinese instruments making tradition imprinted in any way on the Arabic tradition ?

S.M.R. Ansari : According to Samrāt Siddhānta by Jagannāth Dhāt al-Shuᶜbatayan (triquetrum) had been improved and replaced by Rām Yantra.

I may refer to Prof. Hartner's article on this topic.

S.N.Sen and S.R.Sarma (first row from left),
J.E.Kennedy and Mrs.J.E.Kennedy(second row from left)

4.12 A MOORISH ASTROLABE FROM GRANADA

Maria G. Firneis
Astronomical Institute, University of Vienna
Türkenschanzstr.17 ,A-1180 Vienna, Austria

In the course of a systematic search in the "Technical Museum" of Vienna for instruments of astronomical importance, a Moorish astrolabe with Cufic lettering was discovered by the present author. Being labelled as "sun-moon-dial" its correct function had not been recognized before or forgotten. The device came into the museum's possession in April 1937 through exchange with a theodolite from the Jesuit College at Kalksburg where the astrolabe had been kept after it remained in Vienna during one of the two Turkish sieges of the city (probably in 1683).
The brass instrument has a diameter of 15,7 cm and is well equipped with 8 ṣafīḥas (= tympans). Its conservation state is rather good,even the original red and gold ʿilāqa (= cord) is in place. The specimen is of undated planispheric northern type (shamālī musaṭṭaḥ). It shows the typical small kursī (= throne) of occidental instruments, cast in one piece with the ḥajra (= limb) and umm (= mater) forming the main body of the astrolabe. This fact sets it apart from instruments of European origin where the kursī sometimes was fastened with screws to the main body (W. Morley, reprinted in R. Gunther, 1976).
While the front side displays a 360 degree division on the limb, inscribed in quadrants, the umm only shows the tropic of Capricorn, the equator and the circle of the tropic of Cancer. On the rear side, besides another subdivision into degrees, the zodiacal signs are marked also by their astrological symbols . Also a solar calendar circle is shown. Some of the months have been lettered later in Latin writing, most probably indicating Provençalic language (derived from 'OTTO' as shortening for October). 1st of Aries coincides with March 13th placing the object in the first half of the 13th century if this fact can be attributed to the actual date of fabrication. However the instrument may have been produced later, if only an earlier design had merely been copied. Inside the innermost circle with 28 subdivisions the shadow scales for direct and inverted shadow are given in 4 x 3 degrees for terrestrial use. One of the diopters of the ʿiḍāda (the alidade) is missing.

The ʿankabūt (= rete) gives 13 star names inside the ecliptic and 15 outside, a list of which is provided in Table 1.

The 8 ṣafīḥas, each with a diameter of 13,8 cm and a thickness of 0,25 mm (only plate number 4 measures 0,4 mm) are labelled according to the cities and their latitudes for which they are constructed. Somebody has scratched

Table 1. STAR-IDENTIFICATION ON THE RETE

Nr.	Stars inside the zodiac		Nr. according to Kunitzsch (1959)
1	jaḥfala	ε Pegasi	58
2	ṭāʾir	α Aquilae	54
3	ḥawwāʾ	α Ophiuchi	51
4	ʿaṭfat al-ḥayya	δ Serpentis	-
5	mankib al-faras	β Pegasi	62
6	al-ridf	α Cygni	56
7	al-wāqiʿ	α Lyrae	53
8	al-fakka	α Coronae Borealis	45
9	al-ramiḥ	α Bootis	41
10	al-dubb	μ Ursae Majoris	-
11	wasaṭ al-saraṭān	ι Ursae Majoris(?)	-
12	al-ʿayyūq	α Aurigae	20
13	raʾs al-ghūl	β Persei	14
Nr.	**Stars outside the zodiac**		**Nr. according to Kunitzsch (1959)**
1	dhanab al-jady	δ Capricorni	59
2	dhanab qayṭūs	β Ceti	8
3	baṭn qayṭūs	ζ Ceti	6
4	al-jadhmāʾ	α Ceti	13
5	rijl al-jawzāʾ	β Orionis	19
6	ʿabūr	α Canis Majoris	23
7	raʾs shujāʿ	σ Hydrae	-
8	shujāʿ	α Hydrae	29
9	al-ghurāb	γ Corvi	36
10	al-aʿzal	α Virginis	39
11	qalb al-ʿaqrab	α Scorpii	48
12	dabarān	α Tauri	18
13	mankib	α Orionis	22
14	ghumayṣāʾ	α Canis Minoris	25
15	zubānā	α Cancri	-

their Latin names into the lower parts of the corresponding plates. The numbers of the tympans have been crudely cast into the plates, probably when the instrument was registered as museum specimen.

Plate 1a) carries the following inscription: Latitude of Al-Jazīra and the whole province, its latitude is 39°30' (this neither fits to Algiers nor Algeciras).

1b) On the rear side the inscription reads: for the latitude of Merida and the whole province, latitude 42°30'. Here European numerals are written to explain the Cufic ones. Both sides show, besides the usual azimuth and almucantar lines of sudsi - type (with 15 lines marking 6 by 6 degrees) also the lines of the unequal hours plus the information of midday and afternoon prayer times as dotted lines.

Plate 2a) shows the almucantars below the ufq (= horizon) as also known for the Great Astrolabe of Jaipur (V. Nath Sharma, 1984) but for a high geographical latitude. The inscription reads: Longitude of the stars and their latitude.

2b) gives the information that it is constructed for 37°30' northern latitude (which is the value for Granada) and contains astrological information showing the 12 celestial houses divided into 36 parts, consecrated to the 36 decans.

Plate 3a) also shows the celestial house lines for the latitude of 37°30' while the text reads: according to the layout of Hermes (Trismegistos, see also J. Samso, 1973). The ordinal numbers start from the horizon line marked "East" on the left side of the plate and run counterclockwise.

3b) is part of a tablet of horizons showing only one horizon line which in accordance with the geometry (H. Michel, 1947) was recomputed by the author to give 44°27'36". Possibly the tympan was carved by its last Turkish user and could have served with tolerable accuracy for the latitude of Vienna too.

Plate 4a) again states " for the latitude of Granada and the whole province, latitude: 37° 30'." Temporal hours and prayer lines are given also. From the overabundance of latitude 37°30' it can be deduced, that the instrument most probably was produced in Granada.

4b) is a tablet of horizons for all latitudes of the typical occidental style according to the Andalusian astronomers Alī b. Jalaf and Arzaquiel, stating " for all latitudes".

Plate 5a) is " for the latitude of Mālaqa, 37° " (which would be incorrect).

5b) is " for the latitude of Almeria, 36° " (which also is incorrect).

Plate 6a) " For the latitude of Mecca, 21°30' ."

6b) " For the latitude of Aleppo, 35° ."

Plate 7a) " For the latitude of Cufa, 31°30'. "

7b) " For the latitude of Sabta (= Ceuta) , 35°30'. "

Plate 8a) " For the latitude of Fas (= Fez), 33°30'. "

8b) " For the latitude of Marrakesh (Morocco), 30°30'."

As stereographic projection is used for the conception of an astrolabe, the linear distance m between the zenith and the north pole(= center of each tympan) measured in units of the radius r of the equator circle, can be used to estimate the actual geographic latitude for which the tympan was laid out according to the formula:

$$\varphi = 90^{\circ} - 2 . \arctan \left(\frac{m}{r}\right) \quad . \qquad (1)$$

The root mean square error $\Delta\varphi$ for the reconstruction of φ for each plate can be obtained from:

$$\Delta\varphi = \frac{360^{\circ} \ . \ \Delta x}{\pi \sqrt{1 + \left(\frac{m}{r}\right)^{2}}} \qquad (2)$$

where $\Delta x = 0.01$ cm is the precision of the actual measurements. Having reconstructed the latitudes from the tympans, some were compared to modern values to get an idea of their internal accuracy.

Table 2. COMPARISON Of RECONSTRUCTED LATITUDES AND ERRORS

Tympan Nr.	Latitude inscribed	Latitude reconstructed	rms error $\Delta\varphi$
1 front	39°,5	39°,30	1°,03
1 rear	42	41,17	1,04
2 front	-	65,98	1,12
2 rear	37,5	38,30	1,03
3 front	37,5	-	-
3 rear	-	44,46	1,05
4 front	37,5	36,42	1,02
4 rear	-	-	-
5 front	37	35,90	1,02
5 rear	36	34,62	1,01
6 front	21,5	20,40	0,94
6 rear	35	35,38	1,02
7 front	31,5	31,85	1,00
7 rear	35,5	33,98	1,01
8 front	33,5	32,10	1,00
8 rear	30,5	29,40	0,99

The error of the reconstructed latitudes versus the inscribed ones shows the limited precision which was obtained by the astrolabist in manufacturing the instrument. No indication of the craftmanship is mentioned and nothing is known how it came into the possession of the Turks besieging Vienna and why such a precious device was left back by its proprietor. The author owes a hint to J. Samso (1985) that there might be a connection of this astrolabe to Hasan b. Muhammad b. Baso who was chief of the time-reckoning service at the grand mosque of Granada and died 716/ 1316 - 1317.

Acknowledgements:

I am greatly indebted to professor P. Kunitzsch who suggested several improvements to the correct identification of the Cufic star table and to professor J. Samso who helped me with the proper reading of the Cufic inscriptions. I am grateful to both of them for several valuable discussions.

Fig. 1: Backside of the Moorish astrolabe.

References:

Gunther, R. (1976). The Astrolabes of the World, London, The Holland Press.
Kunitzsch, P. (1959). Arabische Sternnamen in Europa, Wiesbaden, Otto Harrassowitz.
Michel, H. (1947). Traité de l'Astrolabe, Paris, Gauthier-Villars.
Nath Sharma, V. (1984). The Great Astrolabe of Jaipur and its Sister Unit, Archaeoastronomy, Nr. 7, Vol 15, S126.
Samso, J. (1973). A propos de quelques manuscrits astronomiques des bibliothèques de Tunis: Contribution a une ètude de l'astrolabe dans l'Espange Musulmane, II. Colloquio Hispano-Tunecino, 171, Madrid.
Samso, J- (1985). Private communication.

DISCUSSION

M.A.N.Mohammed : What was the zero longitude on these astrolabe ?

Maria Firneis : It was not measured by now.

S.N.Sen : Does the astrolabe have latitude and longitude of places inscribed in the mother ? Have you measured the coordinates of the stars inscribed in the ahkābut from the ṣafiha mīzān al-ankabūt! and from there date of the astrolabe.

Maria Firneis : These coordinates have not been measured, but may not be correct indicators for dating, as astrolabists quite often copied older instruments. Also the positions on the 'Ahkābut' sometimes are bent, a fact that makes dating even more uncertain.

4.13 ASTRONOMICAL EFFORTS OF SAWAI JAI SINGH - A REVIEW

Virendra Nath Sharma
Department of Physics and Astronomy, University of Wisconsin
Fox Valley, Menasha, WI 54952
USA

Abstract: Sawai Jai Singh, the statesman astronomer of 18th century India, designed instruments, built observatories, prepared Zīj, and sent a fact-finding scientific mission to Europe. His high-precision instruments were designed to measure time and angles with accuracies of ± 2 second, and ±1' of arc respectively. The Ṣaṣṭhāmsa, a meridian dial with aperture, can still measure angles with precision of ± 1′ of arc. In the age of Newton and Flamsteed, Jai Singh and his associates remained medieval, in the tradition of Ulugh Beg, and did not initiate the new age of astronomy in the country. A complex interaction of poor communications, religious taboos, theological beliefs, national rivalries and plain simple human shortcomings are to be blamed for the failing.

INTRODUCTION

Sawai Jai Singh (1688-1743), the statesman astronomer of 18th century India, undertook the "task (of revising planetary tables), which during a long period of time...since Mirza Ulugh Beg, no one had paid attention to" (See Reference *Zīj Jadīd Muhammad Shāhī* f.1). He constructed five observatories in different cities of north India and, after a considerable amount of labour of seven years or more, had a Zīj prepared. Jai Singh's interests were varied, he collected books, had astronomical texts translated into Sanskrit, and sent a fact-finding scientific mission to Europe.

The object of this paper is to review Jai Singh's diversified efforts to rejuvenate astronomy in his country, and to assess him as an astronomer.

ASTRONOMICAL INSTRUMENTS

Masonry instruments

Jai Singh had been interested in instrumentation long before he decided to erect observatories, and as such, he collected literature on the subject. In 1716, he purchased two books on Turīya Yantra(File No.424/1, Jaipur records, Rajasthan State Archives). He initiated his ambitious program of observing the heavens with instruments of brass constructed according to Persian-Arabic school of astronomy (See Reference *Zīj Jadīd Muhammed Shāhī*,f.1). However, the metal instruments did not measure up to his expectations, and he noted with disappointment that their axes soon wore down, displacing the centre and shifting the planes of reference. A reason for Jai Singh's difficulty could have been the fact that the technology of fabricating large metal instruments had not yet been developed in India. If the surviving specimens at the Jaipur Jantar Mantar are indicative of his attempts to construct metal instruments, it is obvious that his early instruments were too heavy, with no attention paid to eliminate the excessive weight. They were simply magnified versions of the small medieval instruments of the East that may still be seen in museums around the globe.

Jai Singh discarded his metal instruments in favour of ones of stone and masonry that he himself designed. De Bois, an eyewitness of his efforts, writes that the Raja prepared the models with his own hands (De Bois). Altogether, Jai Singh constructed 13 different types of instruments, ranging from a few cm to 24 meters in height, for his observations at Delhi, Jaipur, Varanasi, Ujjain and Mathura (Sharma,V.N. 1987; Kaye 1918). In 1981-82, the author spent several months studying Jai Singh's instruments at these locations, and noted that the instruments may be classified into three main categories, namely: 1. instruments for laymen, 2. low-precision instruments, and 3. the high-precision instruments. The instruments such as Dhruva Paṭṭikā and Narīvalaya have been clearly designed with a layman in mind, and are of little value to a researcher. The instruments such as Jai Prakāśa and Rāma Yantra, on the other hand, belong to the medium-precision category.

Jai Prakāśa is a multipurpose device and has a varying degree of accuracy. The Jaipur instrument, for instance, measures time with an uncertainty of ±1/2 to ±1 min. and the zenith distance and declination both with uncertainty of ±3' of arc. The uncertainty in the measurement of azimuth and right ascension could be anywhere from ±3' to ±1 deg.

Despite the fact that Jagannath lauded the Jai Prakāśa as the finest of instruments, its utility is rather limited for the kind of precision Jai Singh had in mind (Sharma, R.S. 1967). The instrument is a good teaching tool nonetheless, displaying the relationship between the local and the equatorial systems of coordinates. Its ability to indicate the approach of a sign on the meridian is quite valuable to a Hindu astrologer.

The Rāma Yantra, another of his low-precision devices, is most sensitive near the 45^0 marks around the base line of its vertical walls. The theoretical accuracy of the Jaipur yantra is ±1' or arc for the angles of 40^0- 50^0. However, due to the finite width of the penumbra associated with the shadow of the pillar, the accuracies of this order are not possible in practice. And as such, a precision of ±6' is the most that one can expect with this instrument[1]. For the zenith angle readings, the accuracies deteriorate at a rapid rate to $\pm 1^0$ or worse, as the angle approaches zero.

The Samrāṭa Yantra and the Ṣaṣṭhāṃsa both belong to the third category, that of high-precision instruments, and are the two most sensitive instruments indeed. The Samrāṭa of Jaipur is apparently designed to measure time with precision of ±2 second, right ascension ±15" of arc, and declination ±1' of arc. With this instrument, an accomplished observer can measure time with at least ±3 second, and the two angles with ±1' pf arc accuracies, provided the instrument is constructed properly. However, the instruments of Jaipur and Delhi both suffer from a number of constructional defects. As a result, the time measurement accuracy for the Jaipur instrument ranges between ±10 and ±90 seconds, depending on the section of the instrument used. Similarly, the uncertainty in declination measurements could be anywhere from ±1' to ±15' of arc. With his Samrāṭas Jai Singh seems to have measured time with a greater accuracy than the instrument of Jaipur is capable of today.

The Ṣaṣṭhāṃsa of Jaipur is the only sensitive instrument of Jai Singh's that still maintains its intended accuracy of ±1' of arc. In 1981-82, the author took numerous readings with this instrument, and after becoming familiar with the device, his readings began to fall within the intended uncertainty. On December 23, 1981, for example, he measured the apparent declination of the meridian Sun as 23^0:25'±1'. This result compares well with the calculated value of 23^0:26' for the day.

Jai Singh did not stop after having designed these two high-precision instruments, but continued on with his search for better ones. Accordingly, he sent a delegation to Europe in 1728 with one of its goals being to learn about the old and the new instruments for astronomical observations (*Gazeta* 1). He apparently intended to incorporate these instruments into his Jaipur observatory which, according to De Bois, was still being expanded at the time (De Bois).

The questions that the Raja addresses to Boudier also reflect his continued interest in instrumentation. "How is the longitude of the moon observed in its off meridian position, and with what instrument?" he wrote to Boudier in 1730-31 (*Lettres*, pp.610-11)

THE TELESCOPE

Though Jai Singh's instruments are nontelescopic, he was aware of the telescope and to some extent, of its potential as well. He had bought one at a cost of Rs.100. for his personal library (*Tozis* 3) In his Zīj-i Muhammad Shāhī, while introducing a chapter,"The visibility of Moon", he states, "These rules are for naked-eye observations only, although the telescope is now being made in the country. The telescope enables one to see bright stars in broad daylight also--say around the noon hour. It also enables one to see the moon when there is hardly any light in it, or when its face is totally dark and invisible... The planet Saturn (through a telescope) appears oval in shape, an oval whose lower half is larger than the upper. Around the planet Jupiter there revolve four bright stars. On the face of the Sun there are spots, and the Sun rotates once on its axis within a period of one year" (See, *Zīj Jadīd Muhammad Shāhī*, f.189). He goes on to add, "Since the telescope is not readily available to an average person, we are going to base our rules of computations for the naked eye only" (*Zīj Jadīd Muhammad Shāhī*, f.189).

Thus the telescope was indeed available to Jai Singh and his astronomers, and they did use it to some extent at their observatories. In the astronomical endeavors of Jai Singh, it is not the telescope but the "telescopic sight" that is missing. His instruments are not designed to incorporate the telescopic sight. The author believes that the telescopic sight, which had been accepted only decades earlier by the astronomers of Europe, did not reach Jai Singh in time.

Jai Singh's statement regarding the shape of Saturn suggests that the telescope available to him were of inferior quality. His erroneous statement about the rotational period of the Sun indicates that at his observatories there was no sustained program of observing in general with the telescope. His lack of interest in telescopic observation, and that of his astronomers as well, is further indicated by the fact that Jagannath does not include the telescope among the yantras for an observatory (Sharma, R.S. 1967, *Yantrādhyāya*).

THE HINDU, MUSLIM AND EUROPEAN ASSISTANTS

Jai Singh's early training as an astronomer had been under Hindu pundits, and they remained the mainstay of his program until the very end. "Brahmins observe day and night at Jaipur", Boudier wrote (*Lettres*, pp 307-308). In 1734 there were 20 Hindu astronomers employed on daily wages at the observatory of Jaipur alone (*Tozis* 1). Keval Ram and Jagannath were his principal Hindu *Jyotiśīs*.

Jai Singh's interest in Islamic and European astronomy, and his patronizing of Muslim *nujūmīs* and *firangīs* developed somewhat later, i.e., after 1715. An inventory of his library taken in 1715 shows Sanskrit books only (File No. 424/1, Jaipur records, Rajasthan State Archives). However, soon thereafter, Jai Singh started collecting Persian-Arabic works, and began patronizing the predominantly Muslim astronomers of the Persian-Arabic school. The involvement of Muslim astronomers in his program peaked around 1725 and then tapered off until the end of his career (*Dastur Kaumvār*). Jai Singh's most decorated *nujūmī* was Dayanat Khan (Dastur Kaumvār, **19**, p.563). The gifts and honors received by Dayanat Khan from the Raja suggest that he remained associated with him for over two decades. In 1718, he received his very first gift of Rs.500 from the Raja. The final record of him receiving a gift is in 1739.

Abdul Khair, alias Khair Allah, is said to have been one of his influential consultants (Brindāban). However, the author did not see any record of gifts or honors received by Khair Allah in the Rajasthan State Archives. The Archives mention a dozen or so other *nujūmīs*, however, receiving gifts from the Raja. It is reasonable to assume that a team of these astronomers wrote the Zīj-i Muhammad Shāhī under the direction of Jai Singh.

As the involvement of Muslim *nujūmīs* in his astronomical activities lessened, the involvement of Europeans-primarily of Catholic faith-increased, reaching a peak around 1733 (*Dastur Kaumvār*, 18, 20; *Tozis* 2). The Europeans played the role of conveyors of European knowledge to the Raja. Accordingly, it was they who led a delegation to Europe, procured texts and instruments, translated de La Hire's tables, and carried out mathematical computations. However, the knowledge the Europeans brought to Jaipur had already become outdated in Europe, for it included neither the theories of Galileo, Kepler or Newton; nor the instruments such as the sextant employed by Flamsteed (Sharma,V.N. 1987; Kaye 1918).

DELEGATION TO EUROPE

In 1727-28, Jai Singh dispatched a scientific delegation to Europe after learning that "the business of observatory was being carried on in Europe". The delegation, the very first of its kind, led by Fr. Figuerado, reached Portugal in January of 1729 (*Gazata* 2). The Lisbon Gazette of March 10, 1729, reports that the delegation had come to the country to resolve questions regarding the astronomical tables used in Portugal and India, and to learn about the new and old instruments in astronomy (*Gazeta* 1).

Although the news of the delegation's arrival in Lisbon was duly published in the Paris Gazette, there is no mention of its arrival in Paris. The London papers are also silent about the delegation visiting their city. Apparently, the delegation never travelled to those places where the most advanced work on the subject was being carried out at that time. Surprisingly, the records of Coimbra University, the only institution of higher learning in Portugal at the time, say nothing about the delegation reaching their campus either.

The delegation returned in 1730 with some instruments, books on mathematics, and astronomical tables-including the one by de La Hire published in 1702. It is interesting to note that the delegation did not bring back the tables of Flamsteed published in England, which were the most accurate ones by then. In retrospect, the delegation produced very little.

ZĪJ-I MUHAMMAD SHĀHĪ

Jai Singh's stated objective had been to produce a set of improved astronomical tables. The tables, drawn up by a team of observers and *nujūmīs*, as mentioned earlier, and completed sometime between 1727 and 1735, were dedicated to the reigning emperor Muhammad Shah. The tables are, therefore, called Zīj-i Muhammad Shāhī--the astronomical tables of Muhammad shah. Zīj-i Muhammad Shāhī, henceforth abbreviated as ZMS, is a 400-page long traditional work on astronomy, of which a dozen copies survive to this day.

After comparing five different copies of the ZMS, the author has concluded that there were at least two editions of this book[2]. The often-reported British Museum copy belongs to the second edition. Further, there were three commentaries said to have been written on the ZMS[3]. To the world of astronomy at large, the ZMS is of little use; it had been superseded by better tables in Europe even before it was published. However, to the traditional scholar of the country, to whom Western science was out of reach, it remained a valuable resource for generations. The British Resident of Jaipur, J.P. Straton, writing in 1885, pointed out that the ZMS was still being used there at that time (Purohit Hari Narayan Collection).

Recently, it has been reported that "all the tables of the Zīj concerning the Sun, Moon and planets are taken from de La Hire's work... They in no way depend on observations made in India" (Mercier 1984). We have not yet completed our analysis of the ZMS tables. Nonetheless, on the basis of the work done so far, we have reservations regarding the conclusions drawn in the paper.

CONCLUSION

For the sake of rejuvenating astronomy in his country, Jai Singh expanded a great deal of energy as well as his personal fortune, and yet he failed to initiate the new age of astronomy in India. However, it would be unfair to conclude, therefore, that he is solely to blame for this fact. The means of communication were still in primitive stages in the early 18th century, and played a greater role in keeping the Raja ignorant of the contemporary astronomy of Europe than is generally realized. Secondly, a complex interaction of intellectual stagnation, religious taboos, theological beliefs, national rivalries, and the simple human failings of his associates also share the blame to certain extent[4]

Jai Singh's primary interest had been lunar phenomena. And for this, he made the best use of the technology available to him, the technology of building large masonry and stone structure which was highly developed in the country. As pointed out earlier, the great Samrāṭas of Jaipur and Delhi are both capable of achieving a precision of ±1' of arc--the limit for the unaided eye. Further, Jai Singh approached his self-appointed task of updating planetary tables with an open mind, and kept this attitude alive throughout his career. If De Bois can be believed, the Raja was ready to set aside all other tables if a better one were available (De Bois).

Jai Singh's accomplishments were mediaeval in retrospect, but his outlook was quite modern. For him the scientific knowledge had no religion or nationality. His efforts were truly secular; astronomers of all faiths participated in it--a fact that alone is no small compliment to a ruler born in an environment and age of intolerance and bigotry.

ACKNOWLEDGEMENT

The author expresses his thanks and appreciation to Anu Sharma for helping him prepare the manuscript of this paper.

NOTES

1 By superimposing the shadow of a cross-hair on the penumbra one can achieve accuracies better than ±6' of arc. However, it is debatable if Jai Singh's astronomers used such a method.

2 The copies consulted by the author were from the following sources: (1) The British Lib., Ref.1, (2) Raza Lib. Rampur, No.1221, (3) Andhra Pradesh State Archives, Hyderabad, Riyazi 300, (4) Sawai Man Singh II Museum, Jaipur and (5) Azad Lib. Aligarh. The Raza library copy belongs to the first edition.

3 The commentaries were written by: (1) Khairallah See Jaunpuri G.H., (1835). p.579; Khan Gori,S.A.,(1980).

4 For example: Jai Singh's Brahmin pundits would not go to Europe because of a religious taboo against "crossing the ocean". The Europeans (primarily the Jesuits) had rejected Kepler and Newton because of their theological beliefs. The national rivalries in Europe prompted the Portuguese to keep valuable contacts with a powerful prince of the East all to themselves. Some of these points have been further elaborated by the author in: Sharma,V.N. 1982a, 1982b & 1984).

REFERENCES

Brindaban, Safīnā-i Khusgo, f.123, Khudabaksha Lib., Patna.

Dastur Kaumvār, **18, 19 & 20**, Jaipur State, Rajasthan State Archive, Bikaner.

De Bois,Joseph. Introduction to de La Hire's Tabulae astronomicae, ms., Sawai Man Singh II Museum, Jaipur.

Gazeta de Lisboa occidental (in short Gazeta) (1) March **10**,1729, p.80, (2) Jan.**20**, 1729, p.24.

Jām-i Bahadur Khānī by Jaunpuri,G.H., p. 579, Calcutta, 1835; as quoted by Khan Gori,S.A.

Jaunpuri,G.H. (1835). Jām-i Bahādur Khānī, p.579, Calcutta.

Kaye,G.R. (1918). The Astronomical Observatories of Jai Singh,Calcutta. Reprint 1973, Indological Book House, Delhi.

Khan Gori,S.A. (1980). Impact of Modern European Astronomy on Jai Singh, Indian Journal of History of Science, **15**, 50-57.

Lettres édifiantes et curieuses (1843) (in short Lettres) tome deuxieme, Paris.

Mercier,R. (1984). The astronomical Tables of Rajah Jai Singh Sawai, Indian Journal of History of Sciences, **19**(2), 143-171.

Purohit Hari Narayan Collection, No.309/10, Rajasthan Oriental Research Institute, Jaipur, Rajasthan, India.

Samrāṭa Siddhānta of Jagannath Samrata. For ed. see Sharma,R.S. (1967).

Sharma, R.S. (1967). Ed. Samrāṭa Siddhānta of Jagannath Samrat, **2** 1032, Indian Ins. Astro. Sanskrit Research, Reprint 1967, New Delhi.

Sharma, V.N. (1982a). Jai Singh, His European astronomers and the Copernican Revolution, Indian Journal of History of Science, **17** (2), 333-344.

- (1982b). The Impact of Eighteenth Century Jesuit Astronomers on the astronomy of India and China, Indian Journal of History of Science, **17** (2), 345-352.

- (1984). Jesuit Astronomers in Eighteenth Century India, Archives Internationales D'Histoire Des Sciences, **34**, 199-107.

- (1987). The Astronomical Endeavours of Jai Singh, Interaction between Indian and Central Asian Science and Technology in Medieval Times, Indian National Science Academy, New Delhi.

Tozis (1) Bundles Imarat Khana, Jaipur State, V.S. 1791, Rajasthan, State Archives, Bikaner. (2) Daftar Nushkha Punya, V.S. 1770-1797, Jaipur State, Rajasthan State Archives, Bikaner. (3) Pothikhana, Jaipur State, V.S. 1800, Rajasthan State Archives, Bikaner.

Zīj Jadīd Muhammad Shāhī, M.S. Add. 14370: f.1. Dept. Orienteal MSS., The British Library, London, England.

4.14 SKY CALENDARS OF THE INDO-MALAY ARCHIPELAGO

Gene Ammarell
Fiske Planetarium, University of Colorado, U.S.A.

Research conducted over the past several years has revealed a richly diverse astronomical tradition in the Indo-Malay cultural area. I wish, in this paper, to share some of this richness by describing several of the many and diverse observational techniques used by the Indo-Malay peoples to help regulate their agricultural cycles.

Inhabiting mountain sides, river valleys, and coastal plains, the Indo-Malay peoples are faced with a rather unpredictable tropical monsoon climate. In response to this geography and climate they have adopted two distinct types of rice farming: swidden and padi, or dry and wet.

Padi farming is sedentary and relies upon heavy monsoon rains and/or irrigation for the rather large and dependable supply of water required: rice plants must be submerged from the time that they are planted in the nursery through transplanting and until seed is set. From thereon, dry weather is essential for the seed to properly ripen. Padi farming is widely practiced on coastal plains and terraced hillsides by people such as the Javanese, the Sasak of Lombok, and the Malays of Kedah and Perak.

Swidden rice requires less water than padi, but the tropical lateritic soils can provide only limited nourishment under dry cultivation. Usually after 2 years, swidden farmers clear a new area, returning to an old field only after it has remained fallow for 10 to 20 years. Since adequate rainfall is generally available for much of the year in the areas where swidden farming is practiced, the most critical operations are the burn and the dibbling and sowing of the seed. The forest growth must be cut and ready to burn at the height of the dry season, usually no more than a month long, and the planting must occur just as the rain returns. Swidden farming is practiced by the Iban, Kenyah, Kayan, and Maloh, all of northwestern Borneo, among others.

To briefly digress, much of my information has come from sources whose authors have been more or less skilled in observational astronomy. Using a planetarium star projector, I have, in each case, recreated the phenomena described and refined the descriptions in the literature.

For both swidden and padi farmers, nature has long provided dependable markers against which agricultural operations can be timed; recurring celestial phenomena did not go unnoticed. I will now describe examples of the extraordinarily diverse calendrical techniques employed, past and present, by a few of the many cultures of the region. For convenience, I group them using Western astronomical categories as follows: solar gnomons, heliacal apparitions of stars and groups of stars, and lunar-solar and sideral-lunar observations.

A solar gnomon consists of a vertical pole or other similar device that is used to cast a shadow. By measuring the relative length of this shadow each day at local solar noon, one can observe and more or less accurately measure the changing altitude of the sun above the horizon through the year and, thereby, determine the approximate date.

Heliacal (Greek helios: 'sun') apparitions of stars or groups of stars are those which occur at dusk (when the star or stars first become visible in the twilight) or at dawn (when the stars are last seen in the new light). Since the altitudes of stars vary as a function of both the time of night and the day of the year, a certain star or group of stars, when observed at the same time each night, will appear at a given altitude above the eastern or western horizon on one and only one night of the year. Hence the use of any technique or device to measure the altitude of a star or groups of stars as they appear heliacally can provide the observer with the approximate date. The heliacal rising of a star or group of stars occurs on the date that the star or stars are observed just above the eastern horizon at dusk or dawn. Likewise, the heliacal setting of a star or group of stars occurs on the date that the star or stars are observed just above the western horizon at dusk or dawn. The phenomenon that I shall refer to as an "heliacal culmination" occurs on the date that the star or stars under observation reach the height of their nightly climb in the sky (ie. transit the meridian) 1) just as they first appear at dusk or 2) just as they are fading at dawn.

Two distinct types of solar gnomons have been reported. Both measure the length of the sun's shadow at local solar noon to determine the date. The first has been employed by various tribes of the Kenyah. It consists of a precisely measured, permanently secured, plumbed, and decorated vertical hardwood pole (tukar do) and a neatly worked flat measuring stick (aso do), marked with two sets of notches (Figure 1). The first set corresponds to specific parts of the maker's arm and ornaments worn upon it, measured by laying the stick along the radial

Figure 1. Kenyan aso do.

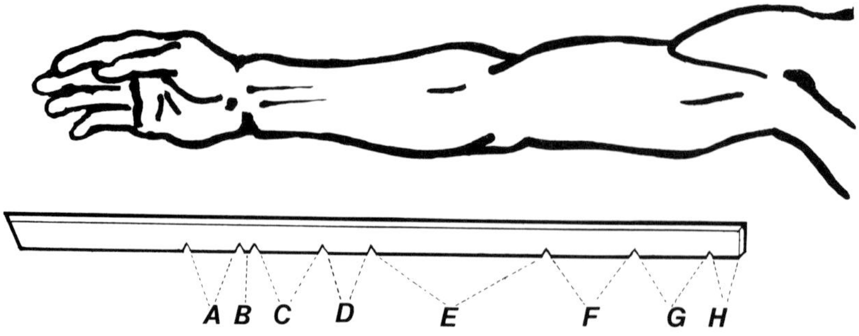

side of the arm, the butt end against the inside of the armpit. The measuring stick is placed at the base of the vertical pole, butt end against the pole and extending southward at the time that the shadow is shortest during the day, local solar noon. On the day that the sun's shadow is longest (the June solstice) a notch is carved to mark its extent on the other edge of the stick. This event indicates that the agricultural season is at hand. The extent of the noonday shadow is then recorded every three days, as a record keeping device. Dates, both favorable and unfavorable, for various operations in rice cultivation, such as clearing, burning, and planting, are determined by the length of the shadow relative to the marks on the stick that correspond to parts of the arm and to the marks made every three days (Hose 1905).

The Kayan, a culture closely associated with the Kenyah, use a similar technique except that the length of the sun's shadow is not measured with a vertical pole in the out-of-doors. Instead, a hole is made in the roof of the weather prophet's room in the longhouse and a measuring stick is securely positioned such that the sunbeam which passes through the hole falls upon the stick at local solar noon (Hose 1905).

On Java a highly accurate gnomon, called a bencet, was in use from about AD 1600 until 1855. A smaller, more portable device than that employed by the Kayan and Kenyah, the bencet divides the year into twelve unequal periods, called mangsa, two of which begin on the days of the zenith sun, when the sun casts no shadow at local solar noon, and another two of which begin on the two solstices, when the sun casts its longest mid-day shadows. At the latitude of Central Java, 7 degrees south, a unique condition exists which is reflected in the bencet. As the illustration shows, when, on the June solstice, the sun stands on the meridian and to the north of the zenith, the shadow length, measured to the south of the base of the vertical pole, is precisely double the length of the shadow, measured to the north, which is cast when the sun, on the December solstice, stands on the meridian south of the zenith. By simply halving the shorter segment and quartering the longer, the Javanese produced a calendar with 12 divisions, divisions which are spatially equal but ranging in duration from 23 to 43 days. Figure 2 shows the bencet and the twelve mangsa with their starting dates and numbers of days.

Figure 2. The Javanese bencet (after Aveni 1981) and mangsa calendar (after Van den Bosch 1980).

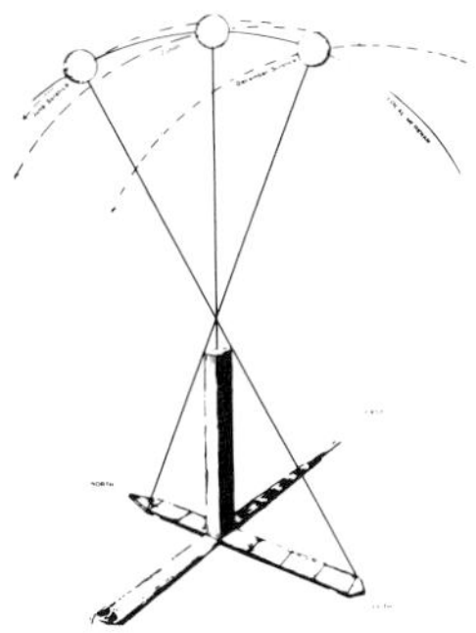

ORDINAL NUMBER	NAME(S)		DURATION	FIRST DAY(S) CIVIL CALENDAR
Ka-1	Kasa		41 Days	21 or 22 Jun
Ka-2	Karo	Kalih	23	1 or 2 Aug
Ka-3	Katelu	Katiga	24	24 or 25 Aug
Ka-4	Kapat	Kasakawan	25	17 or 18 Sep
Ka-5	Kalima	Gangsal	27	12 or 13 Oct
Ka-6	Kanem		43	8 or 9 Nov
Ka-7	Kapitu		43	21 or 22 Dec
Ka-8	Kawolu		26/27	2 or 3 Feb
Ka-9	Kasanga		25	ult Feb/1 Mar
Ka-10	Kasepuluh	Kasadasa	24	25 or 26 Mar
Ka-11	Desta		23	18 or 19 Apr
Ka-12	Sada		41	11 or 12 May

My second category of observational techniques regularly employed by traditional farmers of the region include all of those which involve heliacal (dusk or dawn) apparitions of stars or groups of stars. In a tropical rainforest the horizon may consist of anything from a distant mountain to, more often, some nearby trees; the horizon is, therefore, a rather undependable device against which to sight and measure the positions of astronomical objects. Regardless of location, however, one can usually find a clearing from which much of the sky is visible.

Heliacal risings and settings of stars have been systematically observed by traditional cultures world-wide. From the Indo-Malay region there are references in the literature, too numerous to include here, to the calendrical use of heliacal risings and settings at both dusk and dawn of the stars we know as the Pleiades, Orion and, to a lesser extent, Antares, Scorpius, and Crux. Heliacal culminations (defined above) at both sunrise and sunset of the Pleiades, Orion, and Sirius are noted in the literature. Interestingly, the observation for calendrical purposes of heliacal culminations seems to be unique to peoples of the Indo-Malay Archipelago. The systematic observation of heliacal apparitions of stars by the Iban is described further on; a discussion of such observations made by the Javanese can be found in Ammarell (in press). The heliacal sightings of certain stars at other critical altitudes for calendrical purposes has been recorded throughout the region; three examples follow.

The first technique in this category was practiced by a small Dyak tribe related to the Kenyah-Kayan complex mentioned earlier. Like their neighbors, they were swidden rice farmers but, unlike their neighbors who tracked the sun, this tribe depended upon the stars to fix the date of planting. Each night water was poured into the end of a vertical piece of bamboo in which a line had been inscribed at a certain distance from the open end (Figure 3a). The bamboo pole was then tilted until it pointed toward a certain star (unrecorded) at a certain time of night (also unrecorded), causing some of the water to pour out (Figure 3b). It was then made vertical again and the level of the remaining water noted (Figure 3c). When the level concided with the mark, it was time to plant (Hose and McDougall 1912).

Figure 3. A Dyak bamboo stellar diopter.

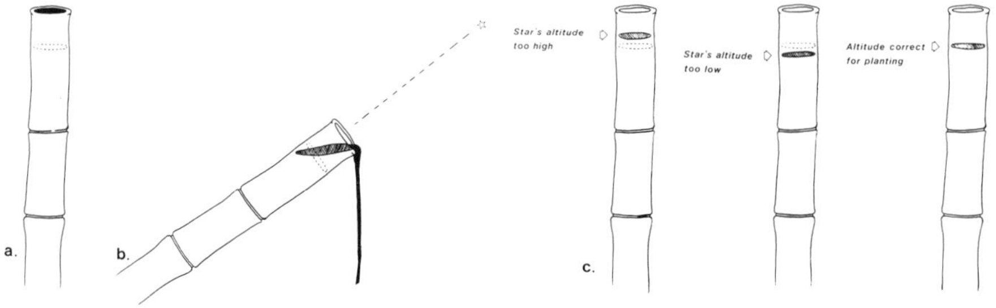

The Maloh, living near but unrelated to the Kenyah-Kayan, used the time between the heliacal culminations at dawn of the Pleiades and Orion (3 - 30 September) to fix the period for planting rice. The time was right for planting, it was pointed out, when a man looked up to see the Pleiades and his hat fell off (King 1985)!

Near Jogyakarta, Central Java, the ritual practitioner (wali puhun) raised his hand toward the East in the direction of Orion at dusk, rice seed in his open palm (Figure 4). On the night that kernals of rice rolled down, it was time to sow seed in the nursery (Van den Bosch 1980). I have fixed the date of this at about 4 January (Ammarell in press).

The systematic observation of the heliacal apparitions of stars by the Iban is described further on, while the observation of both the heliacal apparitions and general altitudes of stars by the Javanese can be found in Ammarell (in press).

Lunar calendars of the region seem to fall into two general categories: lunar-solar and sideral-lunar. Examples of the lunar-solar calendar include the Balinese ceremonial calendar, still in use, and the old Javanese Saka calendar, used from the 8th - 16th Centuries. Both are apparently of a common Hindu origin and are primarily lunar; both employ complex mathematical techniques to provide the intercalary days which periodically synchronize the lunar with the solar year (Covarrubias 1937; Geertz 1973; Van den Bosch 1980). Another lunar calendar which provides intercalary days is indigenous to the island of Savu. But rather than employing direct observations of the sun or mathematical formulae, the Savunese insert an intercalary month, as needed, to synchronize their lunar calendar with the most economically important and reliable occurrence of the year: the second blossoming of the lontar palm, their main source of subsistence (Lebar 1972; Fox 1979).

A second type of lunar calendar is found spread throughout the region, one which uses the heliacal apparitions of stars and groups of stars to determine which month is current. In these cases, it is only important to know which month it is for a few months each year (that is, during

Figure 4. Using rice seed to determine the altitude of Orion (Javanese).

the agricultural season), thereby obviating the need for codified schemes for realigning the lunar with the solar/stellar year. Such 'short' calendars are widely employed in the region (cf Ecklund 1977 & Dove 1981). The Iban sidereal-lunar calendar provides a good example.

The stars play a central role in Iban mythology and agricultural practices. Several Iban stories tell how their knowledge of the stars was handed down to them by their deities (Howell 1963; Freeman 1970; Jensen 1974) and, according to one village headman, "If there were no stars we Iban would be lost, not knowing when to plant; we live by the stars." (Freeman, 1970) The Iban lunar calendar is annually adjusted to the heliacal apparitions of two groups of stars: the Pleiades and the three stars of Orion's belt. That is, when the Pleiades reappear and are visible above the horizon just before sunrise (June 5), the month, taken from new moon to new moon, that is current is called the " 5th Month" (bulan lima). It is during this month that two members of the house go into the forest to seek favorable omens so that the land selected will yield a good crop. This may take from 2 days to a month, but once the omens appear, they return to the longhouse to begin clearing the forest.

If it takes so long for the omens to appear that Orion's belt rises before daybreak (25 June), the people " must make every effort to regain lost time or the crop will be poor." The heliacal rising of Orion at dawn occurs during the 6th month, time to begin clearing the land.

When the Pleiades culminates at dawn (3 September) and Orion is about to do so (26-30 September), it is the 8th month and time to burn and plant. For good yields the burn should occur between the time that the Pleiades and Orion reach the meridian at dawn, usually when the two are in balance or equidistant from the meridian (16 September). Padi sown after Sirius has completed its heliacal culmination at dawn, (October 15) will not mature properly. Planting may carry into the 10th month (October - November), but it must be completed before the moon is full or the crop will fail. At this point the lunar calendar ends: only months 5 - 10 are numbered and fixed while the remaining months vary according to how quickly the crop matures (eg. bulan mantun, the 'weeding month'). The lunar months from November - April are simply not numbered; it is difficult to see the stars during the rainy season and unimportant in any case (Hose 1905; Annonymous 1963; Howell 1963; Freeman 1970; Jensen 1974).

In this brief survey of sky calendars of the Indo-Malay Archipelago I have attempted to share some of the rich diversity of observational techniques employed throughout the region. Of particular interest to the study of indigenous astronomical systems world-wide are the apparently unique calendrical applications of 'heliacal culminations' of groups of stars and the 'short' sidereal-lunar years. The practice and ecological and social context of these calendrical systems calls for further research and will be the subject of a future paper.

REFERENCES

Ammarell, G. (in press). The Planetarium and the Plough: Interpreting Star Calendars of Rural Java.
Annonymous. (1963). In The Sea Dyaks and Other Races of Sarawak, ed. A. Richards, p.82. Kuching: Borneo Literature Bureau.
Aveni, A. (1981). Tropical Archaeoastronomy. Science, 213, no.4504, 169.
Covarrubias, M. (1937). Island of Bali. New York: Knopf, pp.313ff.
Dove, M. (1981). Subsistence Strategies in Rain Forest Swidden Agriculture: The Kantu at Tikul Batu. Stanford University: Doctoral Dissertation, pp.421ff.
Ecklund, J. (1977). Marriage, Seaworms, and Songs: Ritualized Responses to Cultural Change in Sasak Life. Cornell University: Doctoral Dissertation, pp.11113.
Fox, J. (1979). The Ceremonial Systems of Savu. In The Imagination of Reality, ed. A.L. Becker & A. Yengoyan, pp.1526. New Jersey: Ablex.
Freeman, D. (1970). Report on the Iban. London: Athlone Press, pp.171-2.
Geertz, C. (1973). The Interpretation of Cultures. New York: Basic Books, pp.392ff.
Hose, C. (1905). Various Methods of Computing the Time for Planting among the Races of Borneo. J. of the Straits Branch of the Royal Asiatic Society, no.42, 45, 20910.
Hose, C. & McDougall, W. (1912). The Pagan Tribes of Borneo. London: Macmillan, pp.1069.
Howell, W. (1963). In The Sea Dyaks and Other Races of Sarawak, ed. A. Richards, p.967. Kuching: Borneo Literature Bureau.
Jensen, E. (1974). The Iban and Their Religion. Oxford: Clarendon Press, pp.1556 & 1723.
King, V.T. (1985). The Maloh of West Kalimantan. New Jersey: Foris Publications, p.156.
Lebar, F.M. (1972). Ethnic Groups of Insular Southeast Asia. New Haven: Human Relations Area Files.
Van den Bosch, F. (1980). Der javanische Mangsakalender. Bijdragen tot de taal, land en volkenkinde, 136, no.2/3, 25161.

DISCUSSION

E.S.Kennedy : You gave a date in the 17th Century marking the beginning of a certain practice. How did you determine this date ?

Gene Ammarell: By ethnographic records.

ASTRONOMICAL ASPECTS OF "PRANOTOMONGSO" OF THE 19th CENTURY CENTRAL JAVA

Nathanael Daldjoeni
Universitas Satya Wacana, Salatiga, Indonesia

Bambang Hidayat
Bosscha Observatory, ITB, Lembang, Java, Indonesia

It has been shown by Aveny (1981) that the development of indigenous astronomical systems in tropical cultures, whether the motive was religious or practical, centered toward a reference system consisting of zenith and nadir as poles and the horizon as a fundamental reference circle. Such a reference system differs remarkably to the celestial pole-equator (or ecliptic) systems employed by civilizations in temperate zones (also see Brennand, 1896; Stencel, et.al., 1976).

In order to obtain more insight of the view expounded by Aveny (1981) the authors undertook a test case study of the astronomically related time-keeping practice in Java. "Pranotomongso" has been chosen as it is well documented since 1855. According to Daldjoeni (1984) the " pranotomongso" (literaly means the arrangements of seasons) functions well as a practical guide for agricultural activities for the rural peasants in Central Java. The basis of "pranotomongso" is shown schematically in Fig. 1.

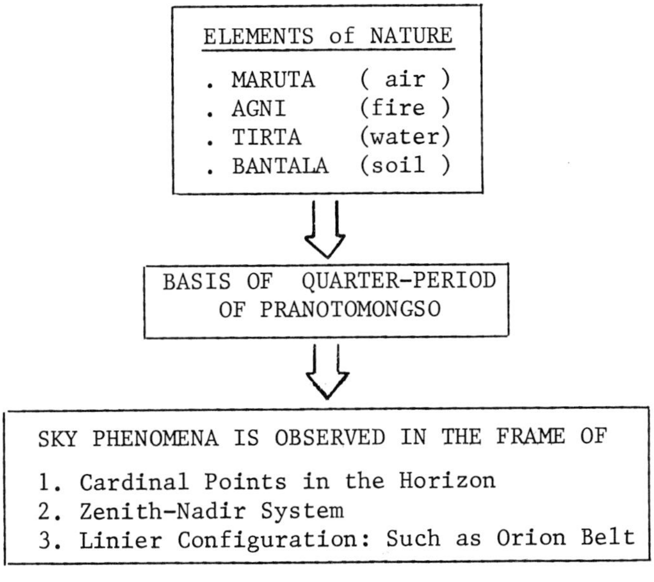

Fig. 1. The basic of "pranotomongso"

The root of the calendar system can be traced back to the old Mataram Kingdom which flourished in the 9th century.

Although according to van Hien (1922, op. cit. Daldjoeni (1978) the use of the time-keeping system started long before the Hindu immigrants arrived from India, the elements of Hindu were later incorporated in it, as can be seen in the names of the stars and star constelations (Maas, 1924; Thiele, 1974). Daldjoeni (1983, 1984) shows the Sanskrit's name of the months which are widely used in the system of "Pranotomongso". It can be seen that there are no extraordinary deviation of orthography, which is Sanskrits.

While the essence of "Pranotomongso" is a tropical year of 365 days, consisting of twelve months of unequal lengths, the calendar reflects also a concept of the world's view of the people in the archipelagoes. Its division into seasons fits the bioclimatological and sociocultural aspects of agricultural activities in the area (of Central Java). Table I shows part of the system related to bioclimatology aspects.

TABLE I. RYTHM of THE SEASON
(Example of character and phenomena of the months)

KASA (KARTIKA) 22/6	. Sotya murca ing embaran (a jewel drops from its setting)	Plants Corn; Cotton; or other Roots Plants
	. Clear Sky; Falling Stars	(Cassava)
DESTA (PADRAWURA) 19/4	. Sotya Sinarawadi (a jewel appears)	Rice Crops; Kapok trees
	"Bird eggs to be hatched"	flowering
SUDDHA (ASUDJI). 12/5	Tirta Sah Saking Sasana (Water gets away from its place)	Festival
	. Cold night, end of rain	

The names of bright stars or apparent clustering of stars that are used to denote the season are mostly, understandably, agriculturally oriented. There are, however, some ambiguities as which stars (or star groupings) are more apropriate for a certain months. A careful examination indicates that:

1. Almost always, bintang Luku (ε, δ and Σ. Orionis; the Belt of Orion) and Bintang Wuluh (Pleaides and, probably, includes Eta Tauri) are

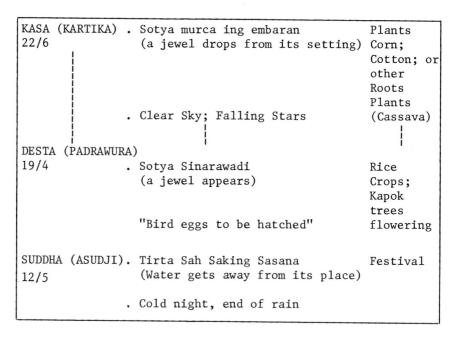

employed to note the beginning of the season. Their positions in the sky, at the times of sunrise or sunset, with respect to the horizon serve as the clock. This strengthens Aveny's view stated above that horizon forms the basic reference. Pannekoek (1961) has also noted that when the Orion's belt is invisible work in the fields ceases and its morning rise (in December) indicates the beginning of the agricultural years.

2. The division of the year into 12 twelve, unequal length, of months can be obtained closely by dividing the apparent orbit of the sun in the sky. For a place in Central Java, which mean geographic latitude is about -7°, the shadow of a sundial casts by midday-sun on June 22 (when the sun is at summer soltice) is twice as long as the shadow in December 21 (when the sun is at winter soltice). The total length of the shadow, which runs exactly north-south, is then divided into six unequal parts which form the first half of the year. The other half of the year is the time taken by the midday sun to trace back the shadow from its extreme north to the extreme south ends.

3. The sign of the month in the night sky must be viewed rather carefully, as the positions of the star groupings may not be exactly given. The most obvious example is the use of the same zodiac symbol for the 2nd Month, (Pusa; 2 August-24 August), and the 12th month (Asudji, Saddha, 12 May-21 June). Both months is under the star grouping which is called "tagih". "Tagih" may represent the whole stretch of the "Milky Way", instead of a single grouping. This is to be contrasted with the use "Bimasakti"--which refers to Sagittarius Cloud, which sets together with the sun at approximately December 22.

Identifications of some Javanese constellations with the commonly used nomenclature in astronomy is shown Table II. Note that the

TABLE II. NAME of STARS

Badak Nyempal	- α Canis Mayoris (Sirius)
Banyak Angrem	- Southern Coalsack
Bima Sakti	- Sagittarius Cloud, probably the whole Southern Milky Way from Norma to Sagittarius
Gotong Mayit	- Head of Scorpio
Hasta	- α, β, Corvus
Kalapa Doyong	- β, δ, π, α, τ, ε, Scorpio
Krittika (Bintang Tujuh)	- Pleiades (May include η Tau)
Lumbung (Gubug Penceng)	- Crux
Waluku	- δ ε Orions (The Orion's Belt) X Orionis may be the bull
Wuluh	- Pleiades (= Bintang Tujuh) but some says: Wulu, which is probably Hyndes
Wulanjar Ngirim	- α and β Centauri

boundary of the groupings are not, as one might have expected, overlapping. There are indications that some names refer to certain stars only.

The financial supports from the Local Organizing Committee, IAU Coll. 91 and travel allowance from the Unesco ROTSEA, in Jakarta, are gratefully acknowledged. We thank "ASTRA" for its part of the support of this study.

References

Aveny, A.F. (1981). "Tropical Archeoastronomy". Science, 213, 161.

Brennand, W. (1896). "Hindu Astronomy". Publ.: Chas, Straker and Sons Ltd., London, 1896, p. 37-81.

Daldjoeni, N. (1983). "Penanggalan Pertanian Jawa Pranatamangsa", in Proyek Javanology, Dept. of Education and Culture, No. 15, 1983 (In Indonesian).

------------- (1984). The Environmentalist, Vol. 4, Suppl. 7, 1984.

Maas, A. (1924). "Sternkunde und Sterndeuterei im malaiischen Archipel". Tijdschrift voor Taal, Land en Volkenkunde, Deel LXIV, Koninklijke Bataviaasch Genootschap van Kunsten en Wetenschappen, Batavia (now Jakarta), 1924.

Pannekoek, A. (1961). A History of Astronomy; Publ. George Allen and Unwin Ltd., London, 1961, p. 22.

Stencel, Robert; Gifford, Fred and Moron; Eleanor. (1976). In "Astronomy and Cosmology at Angkor Wat". Science, 193, 281.

Thiele, Gisela. (1974). "Einiges über das Kalenderwesens Indonesians". Sterne und Weltraum, no. 8/9, 1974, 266.

Van Hien, H.A. (1922). De Javaansche Geestenwerelds. Op. cit. Daldjoeni 1984, p. 16 (Original in Dutch).

USES OF ANCIENT DATA IN MODERN ASTRONOMY

J A. Eddy
National Center for Atmospheric Research, Boulder, CO, 80307, USA

Abstract. Ancient astronomical data are a limited resource that can fill unique needs in modern research. Records from the Orient are of particular value because they were taken more or less continuously, as parts of dynastic and local histories --- often for astrological purposes. We review three problems in modern astronomy that lean heavily on ancient records from the Middle East and Orient: the study of solar variability; studies of the variable rotation of the earth; and studies of the occurrence and physics of novae, supernovae and comets.

INTRODUCTION

What I shall review here are some of the uses that modern astronomy makes of ancient data, and particularly the early records from the Orient.

I think of this as the applied history of astronomy, as opposed to the study of history for its own sake. In this endeavor we attempt to wring useful data from ancient observations --- to rework old mines, with more modern methods, for answers that contemporary data cannot yield. Owen Gingerich has described these activities best as "history in the service of astronomy."

No one has ever thought seriously of starting a journal devoted to the applied history of astronomy, and in today's world that says that the field is very small indeed. The reason it is small is that the corpus of ancient data that is absolutely reliable and adequately documented to be of use is very limited. What is more, all of it has been worked and reworked many times. Practitioners in this field operate much like antique dealers, knowing that the stock of material that makes their trade possible is forever fixed, and that there is nothing one can do to improve either the quality or the quantity of it, regardless of demand. The appearance of anything "new," like the provincial histories, or fang chi, that were unearthed in China by our colleagues there a few years ago, send ripples of excitement through the field (Xu & Jiang 1982).

Like archaeoastronomy, applied history is also a dangerous field, for the temptation is always there, born of hunger or desperation, or national pride, to read more into dusty records than they were ever meant to tell, or to lose sight of the context in which they were made. Nonetheless, there are classes of problems in modern astronomy that lean heavily on old and even ancient records (Stephenson & Clark 1978). The three best known are the study of solar variablitity, resting on naked-eye observations of sunspots and the aurora; studies of the earth's rotation, based on recorded circumstances of solar eclipses; and studies of the physics of novae and supernovae and the mechanics of comets, based on surviving records of these rare phenomena of the sky. I shall briefly review each of these, to point out their promise and some of the present problems involved.

In each of these cases, our need is for continuous and systematic records. At no time and in no place has either criterion been rigorously met for very long; even today. But at times in dynastic China, later in Korea and Japan and much earlier in Babylon, astronomers came close to this ideal. In each of these places these unique, quasi-continuous records were written not as a natural history but as elements of a political one, from naked-eye observations that were made almost exclusively in the name of astrology, and then recorded as a rationalization of what had happened to kings. This unquestionably colors and even edits what was written down. But these accounts, whatever their faults, hold a hundred times more than any early records from the Occident. And it is on these unlikely sources, built on the shaky ground of pseudo-science, that the applied history of astronomy almost wholly leans.

THE SUN

Solar variability is most readily observed through the appearance of sunspots, the largest of which can be seen, by careful observers, with the naked eye (Eddy 1980). Only a handful of these were noted in Europe before the time of the telescope; in the Orient, by comparison, more than 150 are recorded. These naked-eye sunspot reports were accumulated over a span of 17 centuries, yielding an average of about one sunspot per decade, although the number is far from uniform. So small a sample has proven a surprisingly consistent index of the level of solar activity in the past, when tested against contemporaneous indirect and proxy records: the first from historical records of the occurrence of aurorae and the second from measurements of carbon-14 abundance in dated tree-rings, which tell of the solar modulation of galactic cosmic rays (Eddy 1977). The thin record of naked-eye sunspot reports from the Orient can also be used to bolster the case for significant lapses in solar activity in the early telescopic era (Eddy 1976, 1977). For example, a significant drop in naked-eye reports is found to accompany the Maunder Minimum of 1645-1715 (Eddy 1983). It has also been of value in demonstrating the persistence of the 11-year sunspot cycle in pre-telescopic times, through statistical analysis of these naked-eye reports in historical periods when reports of sunspots were sufficiently numerous to withstand the rigors of statistical tests (Clark & Stephenson 1978; Stephenson & Clark 1978; Yunnan Obs. 1979).

Philosophy and even religion have been used to explain why so many more naked-eye sunspots were recorded in the Orient than in the Occident (Needham, 1959). Indeed, the Chinese, unfettered by Christian ideals of divine perfection in the universe may well have been more ready to accept a blemished sun. Meteorological differences have similarly been invoked to explain why these elusive features were most apparent to observers in China, where persistent dust from the Gobi desert may have systematically attenuated the burning brightness of the sun (Willis et al. 1980). The hazy landscapes and dim pastels that characterize Chinese paintings are a tempting confirmation of this convenient conjecture. Moreover, the only records of sunspots from the West --- from Russia in early medieval times --- were made through the haze of forest fires (Bray & Loughhead 1965).

Needham (1959), among others, has suggested that the Oriental astrologers looked at the sun by clever technique: attenuated by reflection in a darkened pool, or perhaps through optical filters. I think this unlikely, given the severe requirements of stability and visual acuity that are needed. It is more likely that most of the 150 Oriental naked-eye sunspot reports come from direct looks stolen from the sun when it was near the horizon, and sufficiently reddened to be viewed without optical aid.

Two enigmas still wrap the Oriental naked-eye sunspot record in something of a mystery: the question of why so few were seen, if an astrological patrol was really maintained, and the riddle of why so small a sample is of any value at all. Stephenson and Clark (1978; Clark & Stephenson 1978) have maintained that sunspots were watched for systematically over the centuries by court astrologers in China, much as these astronomers also patrolled the sky at night. Yet, if we accept the canonical value for the limit of angular resolution of the eye, which is one arc minute, or the experience of modern, naked-eye sunspot observers, then 10,000 times the number that were reported should have been seen in the span of 17 centuries of Chinese dynastic accounts (Eddy 1983).

The vast discrepancy suggests that the 150 sunspots that are recorded come not from a patrol but from occasional or accidental sampling: that large sunspots were there to be seen in all but a few years of each 11-year sunspot cycle but they were only sporadically sampled or noted. Data from this random sampling process --- if that is what it was --- should tell us much more of the observing habits of the Chinese astrologers than of the sun, and this leads to the second enigma: namely that the 150 sunspots seem to be a remarkably good record of the overall level of solar activity on time scales of centuries.

A more trivial unknown is the basis for the curious angular scale that was used by the Chinese and later adopted by Korean court astronomers. Sunspots were reported as dark features on the sun, with sizes described by a variety of mundane objects: "as big as a plum", a date, a peach, a pear,a walnut, or a hen's egg (Stephenson & Clark 1978; Yunnan Obs. 1979). The same comparison objects are used over and over and almost all are edible objects. One is tempted to surmise that the sunspot sightings were made systematically at sunrise, before breakfast, when the astronomers were understandably hungry and thinking about such things. But that is too easy an answer.

What is certain is that these objects were not used as a literal rule-of-thumb, each held at arm's length for comparison against the sun. Any of the objects used for scale would at arm's length cover the half-degree disk of the sun many times over: the seed of an apple would more than suffice to cover the largest sunspot. There may be a clue in the distribution of sizes of the objects used: their dimensions --- the smallest a nut, the largest a pear or a goose's egg -- are in reasonable proportion to the range of relative sizes of visible sunspot groups. If one assumes, as Kevin Yau and I have recently done, that the disk of the sun was thought of as the size of a common, round Chinese washbasin, about a meter in diameter, then nuts and eggs and pears and plums make reasonable, relative measures of large sunspots. If that is so, we may have evidence for the existence of a megalithic washbowl.

THE ROTATION OF THE EARTH

One of the classic problems of celestial mechanics is the determination of the earth's rate of rotation, which is neither constant nor at present wholly predictable. Any astronomer who makes use of the modern Astronomical Almanac is reminded of this fundamental irregularity by the requirement to use Ephemeris Time, which his watch, however expensive, does not exactly keep.

The length of the day is gradually increasing, at a rate of about 2 1/2 seconds per century. The loss of inertial energy is principally through the friction of water against the ocean floor, chiefly on the continental shelves, in the course of tidal motions. Other, non-tidal mechanisms also contribute: one is change in the intrinsic moment of inertia of the planet, brought about by the waxing and waning of polar ice, by the subtle swelling and shrinking of the atmosphere through changes in solar heating, and by shifts of mass in the earth's interior. Another mechanism is that of presumed changes in electromagnetic coupling between the turbulent core of the earth and the mantle above it.

These varying effects on the rate of rotation are noted in accumulated observations made by astronomers with transit telescopes. The long-term, or secular changes of these parameters in the past can be recovered from historical fiducial points that arise from the geometry of solar and lunar eclipses (Stephenson 1982a).

Through such analyses Halley in 1695 noted that the moon's rate of revolution is also decreasing. The associated retreat in distance from the earth is about 5 cm per year which can today be measured directly. The effect is now ascribed to a transfer of the angular momentum lost by the earth, through tidal friction, to the orbit of the moon.

The precise placement and timing of the umbral shadows cast by eclipses is a convolved and sensitive measure of the changing angular momenta of both the earth and the moon. Recovering these circumstances from as many times and as far into the past as possible has been the stock and trade of applied historians (Stephenson & Morrison 1984). Here we profit from the traditional obsession of astronomers and astrologers of all ages with the dramatic phenomena of eclipses. But is has also been the fuel for heated debates, particularly in the more ancient data, where the exact places where recorded eclipses were seen must often be inferred and where times and even dates of occurrence are often vague. Often the crucial distinction between a total and a partial eclipse hangs on the thread of inexact semantics and on the debated translations of ideograms. In the more exact languages of classic or medieval sources, these important distinctions are all too often lost in the mists of allegory.

The best of the older data come from records of solar and lunar eclipses that were compiled by Arab astronomers between the 9th and 11th centuries, A.D. Although clocks were not that accurate, accompanying measurements of the angular altitudes of the moon and sun above the horizon may be used to establish time-of-day with a precision of about 5 minutes in time. These early data were first employed for this purpose by Newcomb in 1878 (Stephenson & Morrison 1984).

More ancient data are found in untimed sightings of total eclipses, mainly from Chinese dynastic histories, and in Babylonian, cuneiform records of the circumstances of lunar eclipses. The Chinese data useful for this purpose are derived from five total and near-total eclipses that were recorded in sufficient detail between 198 B.C. and A.D. 120. The most useful Babylonian records describe about 40 accurately-timed contacts of lunar eclipses between roughly 700 B.C. and 50 B.C. As in the case of sunspots, the useful data describe a minute fraction of the total number of eclipse opportunities that nature provided in this long span.

TRANSIENT PHENOMENA

The third of the major areas of applied history makes use of recorded observations of novae and supernovae and comets.

Bright supernovae were observed by Tycho Brahe and Kepler in 1572, in Cassiopeia, and again in 1604, in Ophiuchus. The first and brighter of these became as bright as Venus and was observed by the famous astronomers for five months; the second for nearly a year. An earlier, more poorly observed event of the same kind was later found in European records from 1006, in Lupus. There have been none since and that is all we would know of these cataclysmic eruptions were it not for early Oriental astronomy and the detailed records kept of the sky in Chinese, Japanese, Korean and Middle Eastern annals.

Chinese and Korean astronomers had observed the same three supernovae that were noted in Europe; in doing so they recorded far more of the course of the earlier two than did their European counterparts. But they also recorded detailed histories of five earlier supernovae, which they called "guest stars": in A.D. 1181, 1054, 393, 386 and 185. In the net that they cast in the sky are also found 67 guest stars of shorter duration, which we now ascribe to galactic novae (Clark & Stephenson 1977).

These Oriental records of galactic novae and supernovae allow a meaningful statistic of the frequency of occurrence of such events; they also provide a basis for deriving their development and expansion with time, which for supernovae are seen today as expanding optical shells and radio sources.

We can only be impressed by the positional detail that was routinely provided in the early Oriental accounts. Tycho's supernova of 1604, for example, is recorded in the following way in the Korean Sillok, in the diary of a single day in a specified month and year:

> "In the first watch of the night a guest star was seen above the stars of T'ien-chiang. It was 11 degrees in Wei lunar mansion and distant 109 degrees from the (celestial) pole. Its form was as large as Venus and its ray emanations were very resplendent. Its color was orange and it was scintillating" (Stephenson & Clark 1978).

The events were scrupulously followed, allowing one to determine, for example, that the supernova seen by the Chinese in Centaurus in A.D. 185 persisted for 20 months.

Comets were no less fixed and followed through the sky, allowing us today to identify every return of Halley's comet since 12 B.C. (Stephenson & Yau 1985). Early positional data secured in this way are especially valuable in studying the long-term motion of the comet. Halley's comet has made several close approaches to the earth during the past two millennia, as in A.D. 837, when it approached within the orbit of the moon. These close encounters of a real kind severely perturb the comet's orbit, and it is impractical to numerically integrate the orbit over the historical period without making use of the early, Oriental accounts.

In some cases the detail given in the Chinese records rivals that of today; as with supernova accounts, it far surpasses what was preserved in the Western world. For the apparition of A.D. 1066, for example, Europe left us the Bayeaux tapestry. The chroniclers of the Sung dynasty, in contrast, gave us this record:

> "On a chi-wei day in the third month of the third year of the Chih-p'ing reign period a comet appeared at the thirteenth lunar mansion. In the morning it was seen at the east measuring about 7 ft., pointing south-west towards the twelfth lunar mansion and reaching Fen-mu. It gradually moved faster towards the east and became concealed when it approached the Sun. Until the evening on a hsin-szu day it appeared at the north-west but without its rays. The comet moved further eastward. Then there was a white vapor about 3 ft. in width penetrating the Tzu-wei and the Pole Star, joining the fourth lunar mansion and with both its head and its tail getting below the horizon.
>
> The comet moved further eastward, passed Wen-ch'ang and Pei-tou and penetrated the sixth lunar mansion. On a jen-wu day the comet retained its rays and measured over 10 ft. in length and 3 ft. in breadth. It was pointing north-east and then it passed Wu-ch'e. The white vapor became branched, stretching horizontally across the heavens, and penetrated Pei-ho, Wu-chu-hou, Hsien-yuan and Wu-ti-tso and Nei-wu-chu-hou with the T'ai-wei.
>
> It reached the first, second, third, and fourth lunar mansions. On a kuei-wei day the comet measured 15 ft. It had a broom-like vapor and resembled a ten-peck measure. From the thirteenth lunar mansion it moved to the 26th lunar mansion passing altogether 14 lunar mansions. The comet and vapor went out of sight after a total of 67 days" (Stephenson & Clark 1978).

For the apparition of 1066 to whom do we owe the greater debt? Is a picture worth a thousand words?

Much of what I have reported on here is based on the work of my British colleague Richard Stephenson. It is also he, with Kevin Yau and Prof. H. Hunger, who has recently sent ripples through the pond of applied history by turning up two very early records of Halley's comet, both of them predating the earliest accepted Chinese account (Stephenson et al. 1985).

The new-found descriptions come from Babylon, where they were incised by an unknown hand on tablets of wet clay in 87 and in 164 B.C. These clay tablets were discovered about a century ago in excavations at Babylon, most of them broken, and were later sold to private collectors. About 2000 of the tablets were later repurchased by the British Museum in London. Some 1200 of these are fragments of astronomical records, mostly datable from about 380 and 40 B.C. (Walker 1985). A number of people have labored long on their translation --- most notably the late A.J. Sachs.

These cuneiform diaries give scattered, day-by-day accounts of lunar and planetary observations and weather conditions in Babylon, with occasional references to comets and other transient phenomena (Stephenson 1982b). Stephenson and his colleagues found three that tell of comets in years when Halley's comet might be expected. There is no question that it is comets that are described; the issue is whether it was Halley's or another. The accomplishment of Stephenson et al. was in deducing the precise positions and dates of inscription, from ancillary information on the double-sided tablets, from the ephemerides of planets and phases of the moon and from a knowledge of Babylonian astronomy. Their paper leaves little doubt that in each case what was described "in the area of Pleiades and Taurus," "one cubit in front of Jupiter," with a "tail 4 cubits long," was the same Halley's comet that we can see, by happenstance, tonight --- from Dehli --- once again near the shining swarm of the Pleiades.

REFERENCES

Bray, R.J. & Loughhead, R.E. (1965). Sunspots. New York, John Wiley and Sons.

Clark, D.H. & Stephenson, F.R. (1977). The Historical Supernovae. Oxford: Pergamon Press.

Clark, D.H. & Stephenson, F.R. (1978). An interpretation of the pre-telescopic sunspot records from the Orient. Q. J. Roy. Astron. Soc., 19, 387-410.

Eddy, J.A. (1976). The Maunder Minimum. Science, 192, 1189-1200.

Eddy, J.A. (1977). The case of the missing sunspots. Scientific American, 236, 80-92.

Eddy, J.A. (1980). The historical record of solar activity. In The Ancient Sun, ed. R.O. Pepin, J.A. Eddy, R.B. Merrill, pp. 119-34. New York: Pergamon Press.

Eddy, J.A. (1983). The Maunder Minimum: a reappraisal. Solar Physics, 89, 195-207.

Needham, J. (1959). Science and Civilization in China. Vol. 3. London: Cambridge University Press.

Stephenson, F.R. (1982a). Historical eclipses. Scientific American, 247(4) 170-83.

Stephenson, F.R. (1982b). The skies of Babylon. New Scientist, 95 (1319), 478-81.

Stephenson, F.R. & Clark, D.H. (1978). Applications of Early Astronomical Records. Bristol: Adam Hilger Ltd.

Stephenson, F.R. & Yau, K.C. (1985). Far Eastern observations of Halley's comet: 240 BC to AD 1368. J. British Interplanetary Society, 38, 195-216.

Stephenson, F.R., Yau, K.K.C & Hunger, H. (1985) Records of Halley's comet on Babylonian tablets. Nature, 314, 587-82.

Stephenson, F.R. & Morrison, L.V. (1984). Long-term changes in the rotation of the earth: 700 B.C. to A.D. 1980. Phil. Trans. R. Soc. London, A313, 47-70.
Walker, C.B.F. (1985). Halley's comet in Babylonia. Nature, 314, 576-7.
Willis, D.M, Easterbrook, M.G. & Stephenson, F.R. (1980). Seasonal variation of oriental sunspot sightings. Nature, 287, 617-9.
Xu Zhen-tao and Jiang Yao-tiao. (1982). The solar activity in the seventeenth century re-assessed in the light of sunspot records in the local gazettes of China. Chin. Astron. Astrophys., 2, 84-90.
Yunnan Observatory, Ancient Sunspot Records Research Group (1979). A re-compilation of our country's records of sunspots through the ages and an inquiry into possible periodicities in their activity. Acta Astron. Sinica, 17, 217-27. (Published in English in Chinese Astron. 1, 347-359, 1977).

4.17 OLD SUNSPOT RECORDS

V.R. Venugopal
Radio Astronomy Centre (TIFR)
P.O.Box 8, Udhagamandalam-643 001
India

Abstract: The works of Eddy (1976) and Clark and Stephenson (1978) on the ancient sunspot records are reviewed and a plea is made for the search for ancient records of astronomical events and phenomena in India.

1 THE ELEVEN YEAR SUNSPOT CYCLE:

Among the large number of solar phenomena that can be observed, the oldest and the best known is the spot activity on the sun. The true discovery of the sunspots as they are described in modern solar physics can be traced to about 1610 when the telescope was used for observing celestial bodies by Galileo. Observations of sunspots in the early seventeenth century have been made by Galileo Galelei, Johannes Fabricius, Father Christoph Scheiner and Johannes Kepler. Galileo and father Scheiner could recognise the movement of the sunspot on the surface from higher to lower latitudes on the sun and correctly conjectured that they were fixed to the sun. Father Scheiner could not only determine the period of the sun's rotation but also noticed that the sunspots changed their latitude in the course of the years. Later in the nineteenth century, following the report of Heinrich Schwabe in 1843 suggesting the existence of an apparent period of approximately ten years between the maxima in the number of sunspots on the visible surface of the sun, Rudolf Wolf made an organised systematic study of the annual means of the relative number of sunspots and found the period to be 11.1 years in 1852. Using scattered data from a host of sources Wolf succeeded in deriving spot numbers back to 1749 and the epochs of maxima and minima back to 1610 when sunspots were observed with the telescope.

The variability of the spottedness is well established and the plot of the annual mean relative sunspot numbers from 1750 to the present seems to indicate a quasi periodicity of roughly 11 years taking on the average about 4 years to rise to a maximum and about 7 years to decrease to a minimum. Carrington, in 1859, discovered the law of latitude drift of sunspots.

2 ANOMALOUS ABSENCE OF SUNSPOTS-DOUBTS ON LAW OF SUNSPOT PERIODICITY:

On the basis of the post 1750 observations of the sunspots one would take for granted that the sunspot cycle of period of approximately 11 years is a fundamental and nonvarying property of the sun. But this does not seem to be true. There have been clear indications from the works of Gustav Sporer and E.W. Maunder in late nineteenth century, who carefully examined the historical records, that during the 70 years between 1645 to 1715 there was a "prolonged sunspot minimum" when the sunspots had almost entirely disappeared. While Maunder's conclusion was considered questionable by many solar astronomers the recent thorough work of Eddy (1976) convincingly establishes that there was a significant decrease in the amplitude of the solar activity cycles from about 1645-1715, an interval coinciding with the reign of Louis XIV in France, when sunspots all but disappeared. This 70 year minimum is called 'Maunder Minimum' by Eddy (1976). Parker (1978) writes: 'The "absurdity" of Maunder's claim that the historical records proved sunspots absent in the late seventeenth century has survived almost to the present day. Only Eddy's recent work has set the topic to rights'. Eddy's plot of the relative deviation in the ^{14}C concentration based on measurements of tree rings and sunspot activity is reproduced in Figure 1. This clearly reveals that ^{14}C record is an indicator of long term variation of solar behaviour. Pretelescopic data on sunspots have been collected by Eddy from published literature on naked eye sunspots. It may be noted that under favourable conditions larger sunspots are easily visible to the naked eye. Actually naked eye sunspots are best seen near sunset when absorption in the atmosphere cuts the normal glare of the sun down far enough that one can look directly at the solar disk.

From this plot Eddy (1976) not only indentified the Maunder minimum 1645-1915 AD but also two other anomalies, namely i) the grand maximum 1100-1250 AD and ii) the Spörer minimum 1460-1550 AD. Clark and Stephenson (1978) on the basis of a catalogue of pretelescopic sunspot records from the orient confirm the existence of 'Maunder minimum' and the 'Sporer Minimum' and suggest the existence of a 'Medieval Minor minimum' in the period 1280-1350 AD. They also point out that these anomalies in sunspot data are supported by the ^{14}C data.

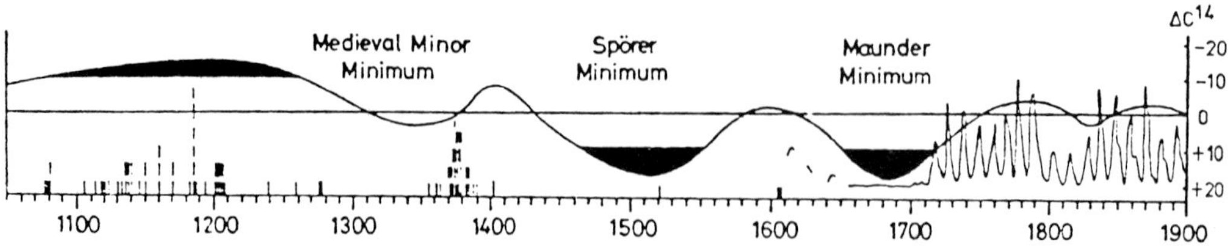

Fig. 1. Reproduction of Fig.8 from Clark and Stephenson (1978)

3 ABSENCE OF ANCIENT SUNSPOT RECORDS FROM THE INDIAN SUBCONTINENT-NEED TO CARRY ON SYSTEMATIC SEARCH.

It is an acknowledged fact that the astronomical tradition in India dates back to the Vedic period, the second millenium BC. There are reasonable evidences to believe that naked eye observations of the Sun, the Moon and the planets and study of their movements and charting of eclipses were carried on by ancient Indian astronomers. There seems to have been an astronomiCcal observatory in Banares called the 'Man Mandir' even as early as the early sixteenth century or even before (Mathur, 1974). In spite of all this it is regrettable as well as ununderstandable that no ancient records describing astronomical events and phenomena such as sunspots, comets, supernovae etc have been found in the Indian sub continent. The reason attributed is that ancient records were lost during the invasions of Muhammad Ghouri and Aurangazeb (Mathur 1974). Even if this were true such destruction was mostly confined to north India and the southern part of India was relatively free from such invasions and destruction. Hence search for historical records of astronomical events such as occurrence of large naked eye sunspots, supernovae like the Crab Nebula of 1054 AD, bright comets etc must be undertaken among the archaelogical inscriptions in ancient temples and old palm leaf manuscripts. Perhaps one can look for any pictographs of such events on ancient rocks and in caves.

Also, H.N. Bhalme and S.K. Jadhav (1984) have studied the patterns of good and bad rainfall over the years in relation to changes in solar activity and have suggested that since 1891 there seems to be a significant tendency for droughts and floods to occur in alternate 11-year sunspot cycles. Since the work of Clark and Stephenson (1978) extends the pretelescopic sunspot records to about 1 AD the study of the link between solar activity and occurrence of floods and droughts in India can be carried out over a longer time scale in order to verify the results of Bhalme and Jadhav. If their results are confirmed over the longer base line of time then this link with solar activity will provide a means of predicting periods of flood or drought risk even though there is, as yet, no clear explanation as to why sunspot changes affect the rainfall.

REFERENCES

Bhalme,H.N. & Jadhav,S.K. (1984). The double (Hale) Sunspot cycle and floods and droughts in India. Weather, **39**, 112.

Clark,David H. & Stephenson,Richard F.(1978). An interpretation of the Pre-Telescopic Sunspot records from the Orient. Q.Jl.R.astr.Soc. **19**, 387.

Eddy,J. A. (1976), The Maunder minimum. Science, **192**, 1189.

Mathur,Krishna D. (1974). Indian astronomy in the era of Copernicus. Nature, **251**, 283.

Parker,E.N. (1978). The New Solar Physics. ed.J.A. Eddy, page 7.

EDITORS' NOTE

Recently Bracewell (1986) and Bracewell & Williams (1986) Have reported interesting result of the existence of two periodicities

~ 314 and 350 Varve years in the sequence of Varves (thickness variations of annual sedimentary layers) from the late-Precambrian formations similar to such periodicities in Sunspot counts and have suggested that from such studies the next solar magnetic cycle could be predicted (References: (1) Bracewell,R.N. (1986). Simulating the Sunspot Cycle. Nature, **323**, 516; (2) Bracewell,R.N. & Williams,G.E. (1986). Hilbert transform analysis of the Elatina Varve record of Solar Activity. Mon. Not. R. astr. Soc., **223**, 457).

4.18 GUEST STARS: HISTORICAL SUPERNOVAE AND REMNANTS

T. Velusamy
Radio astronomy Centre (TIFR)
P.O.Box.8, Udhagamandalam-643 001
India

Supernova explosions occur in massive stars in their final stages of stellar evolution. The explosion leaves behind a collapsed object, usually a neutron star observable as a pulsar, and an expanding outer envelope of the star observable as a supernova remnant (SNR). Because of the large amount of energy in the range of 10^{50}-10^{52} ergs released in the explosion, the energy radiated by the exploded star is equivalent to that of millions of normal stars. Thus a "new star" seems to appear and is easily visible to the naked eye for several months. Astronomers have estimated that in our galaxy such supernova explosions occur once about every 50 years. In ancient times, the occurrence of supernova might have been noticed as appearance of a bright new star which faded away into obscruity after a few months. Thus the "guest stars" in the ancient records are indeed possible occurrences of supernova explosions in the galaxy. The guest star of 1054 A.D., well recorded by Chinese and Japanese astronomers is the best known historic supernova (Brecher et al. 1983; and references therein); today its remnants the Crab Nebula and the pulsar are observable over the entire electromagnetic spectrum from radio to γ-rays. Of the 150 SNRs known in our galaxy, only about 10 have been identified with guest stars of historical supernovae.

Since the invention of telescope no supernova explosion has been observed in our galaxy and our observational knowledge is confined to supernovae occuring in distant external galaxies. For any "observational" information about the supernova explosions in our own galaxy we have to depend only on the records of pretelescopic civilizations. Today we know of nine pretelescopic historical supernovae. The supernovae of AD 1572 and AD 1604 have been well recorded by the famous astronomers Tycho and Kepler. The identifications of the others are based mostly on Chinese records, often confirmed by Korean and Japanese records. Also the ancient literature and historic records of many other civilizations, such as those in ancient Greece, Rome, Arab, Babylon, Europe have been searched for any records of the occurance of guest stars but no evidence has been found so far. However, the Indian literature and records essentially remain unexamined for guest stars.

A detailed account of the historical records examined so far is given by Clark & Stephenson (1977) and in the later reviews by Clark & Stephenson (1982) and Trimble & Clark (1985). As the ancient Chinese believed in the controlling influence of heavens on the destinies of men, in particular kings, professional astrologers and astronomers were appointed to maintain a constant watch of the sky in order to report and interpret any unusual events happening in the sky. Thus the Chinese have official records of astronomical events during various dynasties with almost a continuous record covering nearly 2000 years, as far back from Han Dynasty (202BC-220AD). We now find a wealth of astronomical recordings of eclipses, comets, meteors, planetary conjunctions, new stars etc. Similar records have been maintained by the Koreans and Japanese during the later years. How does one recognise a supernova in these records? The appearance of a new star are recorded typically as: "guest star or visiting star; broom stars or sweeping stars", "ray stars or bushy stars". The appearances of comets can be easily identified from the description of their motion, and the others are potential events of novae and supernovae. In Fig.1 are shown the light curves of Nova Cygni 1975, Type I supernovae observed in external galaxies, and light curve of supernova AD1054 deduced from descriptions in the ancient historical records. Since supernova are brighter than novae by several orders of magnitude and are observable for much longer period, it is possible to distinguish between them. In a supernova the increase to maximum brightness is slower than for a nova and it is possible to detect a supernova before it reaches maximum brightness. A supernova explosion occuring about 2 kpc (~6000 light years) away from the earth will have a peak brightness of about -5mag and can be seen by naked eyes in daytime for about 3 weeks and in night for several months. Further, in a supernovae explosion the outer envelope of the exploded star is ejected out into the surrounding interstellar space at velocities of several thousand km/sec. This expanding envelope, known as supernova remnant (SNR) contains gas and very energetic particles, and the emission from it is observable with radio, optical and X-ray telescopes for thousands of years after the explosion.

After searching through the records available over two thousand years period, Clark & Stephenson (1977) find evidences for occurence of 75 new stars. Of these only about 20 were visible for periods longer than 50 days, and these are possible candidates for historical supernovae. In Tables 1 and 2 we have reproduced from Clark & Stephenson (1977) the list of medium and long duration new stars, and identification of some of the new stars as the historical supernovae and their remnants. For example, the description of the guest star AD1054 (the historical supernova of the Crab Nebula in the Taurus) in the astronomical treatise of the Sung dynasty compiled around AD1345 is given as: "In the first year of the period Chih-ho (July 4) (a guest star) appeared approximately several inches (roughly 0.3 to 0.5 degrees) to the south-east of T'ien Kuan (the star Zeta Tauri). After a year or more it gradually vanished".

In Tables 1 and 2 we have included the guest star of AD1408. Although there is no mention of its duration, its description as reported by Li (1979) suggests that it might have been a supernova: "Reign Ying-le Year 6, winter, month 10, day Geng-chen, at night near the meridian, to the south-east of Niandao, there is a star like a lamp, its colour is yellow, its lusture smooth, it shows up and does not move, it is a

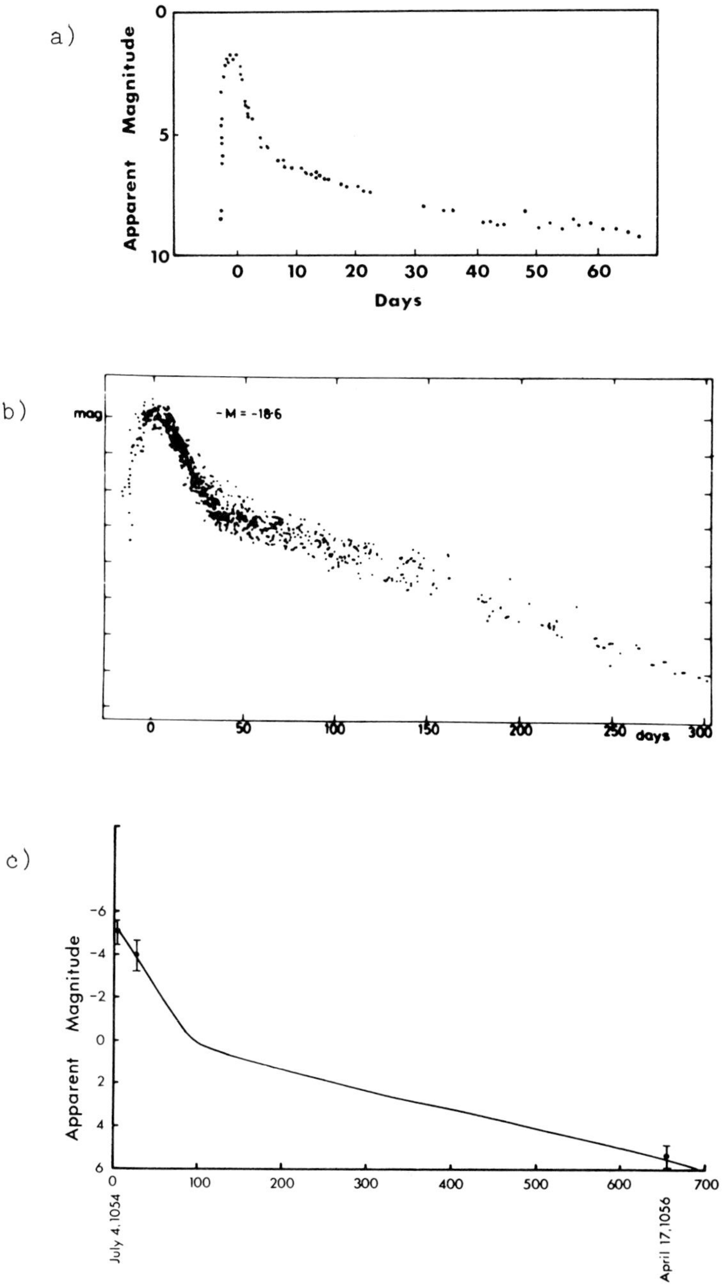

Fig.1. Observed brightness as a function of time,
a) Nova Cygni 1975
b) Type I supernova observed in external galaxies
c) AD1054 Supernova explosion (the Crab Nebula) as deduced from historical records.

Zhou-lo, a virtuous star". Also its location in the sky in the region of the peculiar supernova remnant CTB80 suggests an interesting identification (Strom, Angerhofer & Velusamy, 1980; Wang & Seward, 1984).

TABLE 1
LONG DURATION "NEW STARS"

(1) Date	(2) Recorded in	(3) chinese description	(4) Duration	(5) Approximate Galactic coordinates l	b	(6) Remarks
Mar/Apr 5 BC	China	hui	70 days	30°	-25°	comet/nova?
Sep 27 AD 61	China	K'O	70 days	60	-70	comet/nova?
May 3 AD 64	China	K'O	75 days	290	+55	comet/nova?
Dec/Jan AD 64	China	K'O	48 days	215	+45	nova?
Dec 7 AD 185	China	K'O	20 months	315	- 2	supernova?
Jan 16 AD 247	China	hui	156 days	295	+40	comet/nova?
Mar/Apr AD 369	China	K'O	5 months	-	-	position estimate impossible
Apr/May AD 386	China	K'O	3 months	10	0	possible supernova
Feb/Mar AD 393	China	K'O	8 months	345	0	supernova
Jul/Aug AD 396	China	star	50 days	175	-25	nova?
Nov/Dec AD 402	China	K'O	2 months	240	+60	comet/nova?
May 3 AD 837	China	K'O	75 days	280	+65	nova?
May 1 AD 1006	Arab China, Europe, Japan	K'O	several years	330	+15	supernova
Jul 4 AD 1054	Arab China, Japan	K'O	22 months	185	- 6	supernova
Aug 6 AD 1181	China Japan	K'O	185 days	130	+ 3	supernova
Oct 24 AD 1408	China	star?		69	+ 3	supernova?
Nov 8 AD 1572	China Europe Korea	K'O	16 months	120	+ 1	Tycho's supernova Type I
Nov 28 AD 1592A	Korea	K'O	15 months	150	-70	Mira Ceti?
Nov 30 AD 1592B	Korea	K'O	3 months	125	0	?
Dec 4 AD 1592C	Korea	K'O	4 months	115	0	?
Oct 8 AD 1604B	China Europe Korea	K'O	12 months	5	+ 7	Keplers supernova - Type I

hui broom star; K'O Guest star

Finally, Ashworth (1977) has suggested the only possible historic supernova observed after the advent of modern telescope, the candidate AD1679 (Cas A) in Table 2. This identification is based on Flamsteed's observation of a 6th magnitude star on 16th Aug 1680, designated 3 cassiopeiae in the 1725 Flamsteed catalogue. Later astronomers have dismissed this star as spurious. However its close proximity to the supernovae remnant Cas A strongly suggests that Flamsteed observed the faint supernova long after the explosion during its decay phase.

TABLE 2
THE HISTORICAL SUPERNOVAE AND THEIR REMNANTS

Supernova	Radio remnant	Remarks	Approx distances from sun
AD 185	G315.4-2.3 (RCW 86)	Probable	<2kpc
AD 386	G11.2-0.3	Possible	>5kpc
AD 393	G348.5+0.1 or G348.7+0.3 (CTB37 A and B)	Possible	>6kpc
AD 1006	G327.6+14.5 (PKS 1459-41)	Certain	~1 kpc
AD 1054	G184.6-5.8 (Crab)	Certain	~2 kpc
AD 1181	G130.7+3.1 (3C58)	Probable	~8 kpc
AD 1408	G69+2.7 (CTB80)	Probable	~3 kpc
AD 1572	G120.1+1.4 (Tycho's SN)	Certain	~3 kpc
AD 1604	G4.5+6.8 (Kepler's SN)	Certain	~5 kpc
AD 1679	G111.7-2.1 (Cas A)	Probable	~3 kpc

It may be noted that the identifications given in Table 2 are far from being complete. Very recently Wang et al. (1986) have reported six new identifications of historical supernovae from the Chinese records. Guest stars BC523, BC134, BC48, AD125, AD421, AD1523 have been identified as historic supernovae. Identifications of historical supernova are of great astrophysical importance. Accurate ages of the SNRs obtained from their historic supernovae, are exteremely valuable for better understanding of the physical conditions and processes in the individual remnants. Further, completeness of the historic supernovae will be useful to improve the statistics of the supernovae rate in our galaxy.

REFERENCES

Ashworth,W.B. (1979). Bull.Amer.Astr.Soc. **11**, 660.

Brecher,K., Fesen,R.A., Maran,S.P. and Brandt,J.C. (1983). Observatory, **103**, 106.

Clark,D.H. and Stephenson,F.R.,(1977. Historical Supernova. Oxford: Pergamon.

Clark,D.H. and Stephenson,F.R., (1982).In Supernovae: A survey of Current Research, eds. M.J.Rees and R.J. Stoneham, p.355.

Li, Q.B., (1979). Chinese astronomy. **3**, 315.

Strom, R.G., Angerhofer,P.E. and Velusamy,T. (1980). Nature. **284**, 38.

Wang,Z.R. and Seward,F.D. (1984). Astrophys. J. **285**, 607.

Wang, Z.R., Liu, J.Y., Gorenstein,P. and Zombeck,M.V. (1986).In Highlights of Astronomy, ed. J.B.Swings, **7**, pp. 583-584. Dodrecht: Reidel..

Trimble,V. and Clark,D.H., (1985). Bull. Astr.Soc. India. **13**, 117.

5
Concluding remarks

We must look now to the future prospects. What we do not need is an intensive competition for who did something first, or for who had it right; what we need is to learn about the intellectual framework for these inventions and to learn why discoveries took place when they did.

- O.Gingerich (p.276)

Mary Helen Kennedy and O.Gingerich (first row from left),
A.Moitra and Minati Ray Chaudhuri (second row from left).

5 Concluding remarks: ORIENTAL ASTRONOMY AND THE NATURE OF SCIENCE

Owen Gingerich
Harvard-Smithsonian Center for Astrophysics
Cambridge, MA 02138,
U.S.A.

I have come to this colloquium not as a specialist in Oriental Astronomy, but as a historian of science passionately interested in the fundamental nature of science and its attempts to describe physical reality. A basic question in the philosophy of science concerns the nature of scientific laws: are scientific laws intrinsic to the universe, simply waiting to be discovered? Or are they in part creative inventions of the human mind, our particular way of looking at nature?

Perhaps the best approach to this open problem lies in comparative science. Do independent scientific systems inevitably lead to the same description of nature? As a first step in such an investigation, we must decide if a concept is an independent invention or a transmission from another culture or an earlier epoch. Only if the concepts are truly independent can we begin to compare genuine alternative descriptions, and in fact we cannot be sure that differing systems are even possible until we establish such independence. Olaf Pedersen addressed this question with respect to Greek astronomy, pointing out that if we have identical numerical parameters in two different situations, we can be fairly sure that transmission is involved, whereas it is difficult to be sure if an idea is independent or dependent on a transmission.

Consider the case of the epicyclet device used by Copernicus to eliminate the Ptolemaic equant. An identical construction was adopted by Ibn ash-Shatir nearly two centuries earlier, but no avenue of transmission has been traced. Did Copernicus invent this device independently? I think an independent invention is entirely possible, but transmission is by no means excluded. Even if Copernicus discovered the construction independently, he certainly inherited the same philosophical framework that motivated such an invention, namely, the reconciliation of the mathematical models of Ptolemy with the physical form of the celestial spheres. The example clearly demonstrates the ambiguity of deciding the independence or dependence of scientific ideas.

In any event, there is probably no better place to study comparative science than in Oriental Astronomy. Nevertheless, the field poses formidable problems, foremost being the variety of languages involved. It does no good simply to know Chinese and no other language. We need Chinese scholars who know Sanskrit, Arabic scholars who read Chinese, Indian scholars who command Greek, and so on. I was impressed and envious to watch Dr. Teboul translating questions into Chinese and the answers into English, or Prof. Kennedy likewise speaking in Persian. Despite the significant progress demonstrated in the colloquium, we must recognize that these studies are only in their infancy, primarily because of the language barriers.

In particular, we need a far deeper examination of the transmission problems. E.S. Kennedy graphically described some of the paths between Greek, Indian and Islamic astronomy, but there is a scholarly terra incognita with respect to China. Just before this colloquium I had an opportunity to visit a thousand-year-old mosque in the ancient Chinese capital, Xian. At the end of the Silk Road, Xian clearly provided a gateway for the infusion of Islamic astronomy into China long before the Jesuit astronomer-missionaries of the sixteenth and seventeenth centuries arrived. Yet we know almost nothing of what happened then. Even earlier transmissions occurred, as indicated by the materials now being examined in the Buddhist temples of Japan.

One of the high points of this colloquium came in the exchange following Prof. Nakayama's paper. He had demonstrated the existence, in one of these Buddhist texts, of a method of parabolic interpolation unknown in Indian or Islamic texts, and he pointed to central Asia as a likely source. In the discussion, Dr. Mercier noted that a similar interpolation technique had already been reported from Tibet. It is such connections, individually perhaps minor, that will eventually build up our fabric of understanding concerning the dependence or independence of these scientific cultures, and give us a better possibility to answer whether scientific pictures can be independently formulated and directed in convergent paths.

A recurrent theme of the colloquium has involved the question of observations. One of the distinctive characteristics of Chinese astronomy, pointed out in Prof. Xi's contribution, is the long series of records of unusual celestial phenomena. This stands in striking contrast to India, the Islamic world, and the Latin West. In fact, it is a source of embarrassment that the West can scarcely point to any indigenous observations of the splendid supernova of 1054, the progenitor of the Crab Nebula. However, I think we may well be asking something of the Western chronicles that may be too much to expect. China had a long tradition of writing, printing, and the preparation of official histories with a high regard for omens. The West had neither the advantage of relative stability nor the disadvantage of a stultifying traditional bureaucracy, so the absence of supernova records in the West is an ambiguous deficiency.

There is, furthermore, a qualitative difference between the phenomenological omen observations so abundantly recorded in the Oriental annals, and the positional observations required for the advance of planetary theories. What appears puzzling and even embarrassing to modern astronomers is the great paucity of such measurements in Asia as well as in Europe. This very fact must be telling us something about the different perceptions of our astronomical forebears and their notions of how to extract the secrets of nature, and this may also be revealing something about the uniqueness of modern science.

Several papers have dealt with instruments, which are related to the aforementioned problem of the existence or precision of positional observations. Again, we must be exceedingly careful not to be anachronistic when we discuss instruments of past ages. There were, for example, in Mughal India four generations of Lahore astrolabists, whose instrument designs stayed virtually constant for a century and a half. Now if these astrolabes were actually used for observations, we would expect the less-than-optimal style of the alidade to have been markedly improved over that period. The fact that this did not happen suggests that these instruments were designed for a different purpose than actually measuring celestial altitudes. Quite likely they were in part symbolic, providing, as the late Derek DeSolla Price argued, the opportunity to hold the machinery of the universe in the palm of one's hand. (In a sense today's wrist watches furnish the same connections with the universe at large.)

Not only astrolabes, but even the Jai Singh observatories must be at least partially, if not primarily, seen as cultural and symbolic monuments, and not truly working instruments. One could even remark today on how few telescopes on our college campuses are really observational research tools, and what a large fraction are there for the inspirational opportunity to get in closer touch with the universe. Nevertheless, it seems that at least one of the stone instruments in Delhi was successfully used to determine the latitude and the obliquity of the ecliptic, the use of these parameters in tables giving evidence of observing in the absence of specific extant observations.

In this regard the research reported by Dr. Mercier strikes me as particularly challenging. Using techniques pioneered by R. Billard, he has shown for a variety of Indian tables that the coordinate errors for the planetary positions come to a well-defined minimum at a specific time characteristic of each table. The sheer volume of material forces us to agree that the parameters were generally optimized at the time the tables were made. But is it necessarily true that the parameters were controlled by specific but no longer existing observations? David Pingree has argued that the parameters could have been established from existing Greek tables, and the Indian attempts to force the planets into arbitrary conjunctions at considerably earlier dates would have thrown out the agreement with the actual planetary positions except around the time of the tables. In the absence of any observations, or even records of how they might have been achieved, we must be very cautious in accepting the speculative conclusion that the parameters of the Indian tables were observationally established by the table makers.

Such dating techniques have to be used with great care. For example, Samuel Goldstein has attempted to date the Ptolemaic lunar parameters in a similar fashion, and his interpretation, that the parameters predate Ptolemy by several centuries and were therefore already established from observations of unknown earlier astronomers, are difficult to reconcile with reasonable assumptions about the precision of observational techniques or (equally importantly) the accuracy of whatever coordinate reference frame was used at such an epoch. In a more extreme case, two papers presented here dated the nakshatra (the Indian lunar mansions) to 2400 B.C. on the grounds that only then did the Pleiades rise directly in the east. Can we really demand such precision from the Vedic texts? Can we rationally accept a set of Indian lunar mansions dating that far back in the absence of any corroborative evidence from archaeology? If so, then what about the work of Bradley Schaefer (not reported here) on the Chinese hsiu, where a computer analysis based on various assumptions of symmetry and positions leads to a date before 3000 B.C. for the Chinese system of lunar mansions? I believe it is still an open question about where the lunar mansions originated -- surely a fascinating question regarding transmission or possible independent invention -- but I am sceptical that such dating procedures can by themselves lead us to firm ground. In contrast, I find the results of the Billard-Mercier method entirely reasonable, although I would like to see their findings placed into a convincing context with respect to observational realities.

By probing the character of Oriental Astronomy, the questions of transmission, and the roles of instruments and observations, our colloquium has become a long-overdue pioneering venture in a field that promises high returns for scholarly investment. We must look now to the future prospects. What we do not need is an intensive competition for who did something first, or for who had it right; what we need is to learn about the intellectual framework for these inventions and to learn why discoveries took place when they did. We need to examine closely the roles of innovation versus tradition in the ongoing development of astronomical theory and techniques. To do this we must cultivate skills in languages, we must analyze the texts, and we must search diligently for the evidences of cross cultural transmissions. On such a solid basis we can perhaps begin to illuminate some of the fascinating and fundamental problems about the nature of science itself.

SUBJECT INDEX

MANUSCRIPT INDEX

ASTRONOMERS INDEX

AUTHORS' INDEX